CALIFORNIA TREES

A FIELD GUIDE TO THE NATIVE SPECIES

MATT RITTER and MICHAEL KAUFFMANN

Oak woodland meets redwood forest, Humboldt County

First Edition

9 8 7 6 5 4 3 2 1

Cover design by Backcountry Press

- Front cover *Pinus longaeva* by Matt Ritter
- Back cover photos (top to bottom) *A. negundo* (Mary K. Hanson), *Yucca brevifolia* (Ritter), *Pinus jeffreyi* (Kauffmann), *Sequoiadendron giganteum* (Chris Earle), *Notholithocarpus densiflorus* (Kyle Nessen).

Book layout and cover design by Backcountry Press with AP+MK

All photos by Ritter and Kauffmann except those noted on page 201

Published by Backcountry Press • Kneeland, California

ISBN 978-1-941624-33-3

Library of Congress Control Number: 2025931733

Order this book online: www.backcountrypress.com

BACKCOUNTRY PRESS
Humboldt County, CA

FSC www.fsc.org MIX Paper | Supporting responsible forestry FSC® C005010

Created and Printed in the USA

Table of Contents

Pinus torreyana on Santa Rosa Island, Channel Islands N.P., Santa Barbara Co.

More to Explore

How This Book Works

As humans, we derive meaning from the connections we make in the world and where we focus our attention. Our attention is increasingly focused on screens and the built environment. Connecting with trees, identifying them, and learning more about them can be an entry point into nature and an antidote to the stresses of modern-day life. We wrote this book to help beginners and experts alike become more familiar with the native trees of California—to enable them to enter wild landscapes with a greater understanding of the organisms we share those spaces with. Familiarity with the largest and most dominant plants in wild habitats arouses curiosity about less common ones. Before we know it, we've formed new and meaningful relationships with non-human organisms, and we are living richer, more fulfilling lives full of wonder and appreciation for the gifts of the natural world.

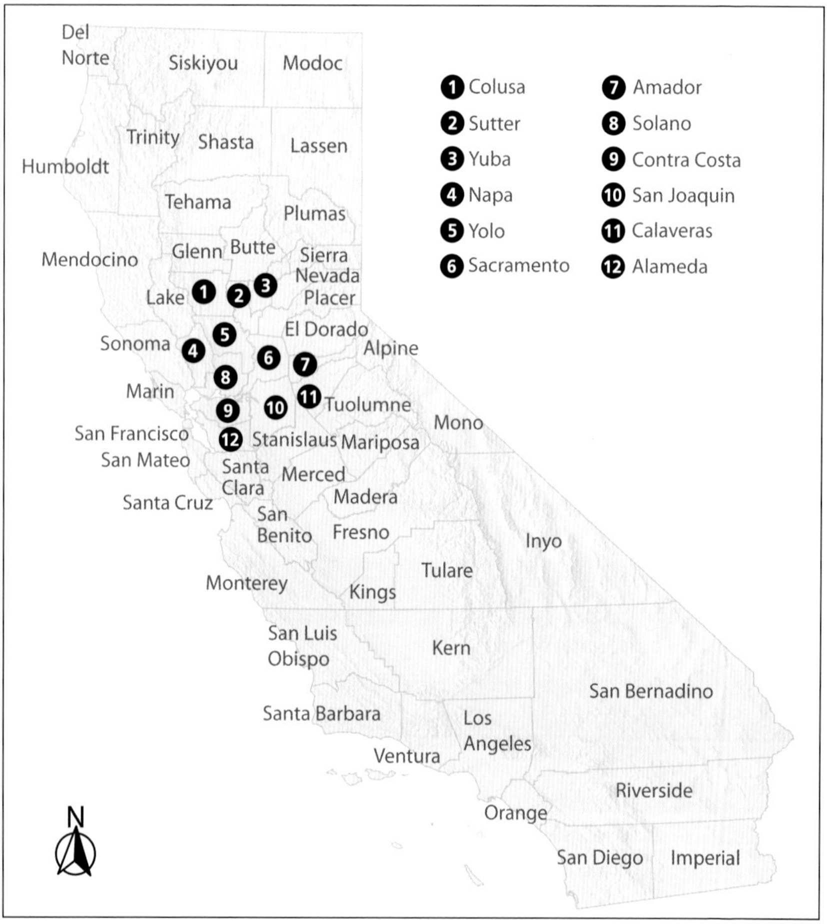

California Counties

We organized the 95 native tree species in this book by their plant families in alphabetical order in two large groups: conifers and flowering trees. Within plant families, we generally listed species by their genus in alphabetical order with a few exceptions. There is introductory text for species-rich groups and for each plant family. There are quite a few tree species that are not native, but have naturalized. Matt Ritter's *A Californian's Guide to the Trees among Us* is a great resource for those species. Also see the treatment of Large and Iconic Shrubs (p. 182) and Non-native Trees (p. 188) at the end of this book.

With the help of Cami Pawlak, we created new range maps from previously published maps, descriptions of known locations, and our experience of many years of wandering California, looking for and observing trees. Although this book and its maps can point you in the right direction, there is no substitute for visiting wild places in California where our native trees thrive. For this reason, we have included our suggestions (on p. 192–199) for the best places to see native trees in different habitats throughout the state.

Each entry in the book describes a single species. Descriptions include a species' range, its height, typical diameter at breast height (DBH — 4.5 feet above the ground) for mature individuals, and characteristics helpful for identification.

Introduction

By Matt Ritter

Trees provide us with our most intimate contact with nature. Their canopies reaching to the sky, their roots held solidly in the soil, they seem to bind the world together and ground us in our place. To be near a large and ancient tree is to feel a sense of awe not easily described on the page or communicated in a photograph. Such trees fill us with a much-needed and often deserved sense of irrelevance and impermanence. Like humans, trees become more interesting as we learn more about them, and our lives are enriched by paying attention to them.

My journey to learn more about trees began many years ago through an interest in wood and furniture making. It has been a 20-year love affair, sometimes filled with heartbreak, as old and magnificent trees I have known have died. My curiosity about these fascinating organisms is continually stoked by new encounters with unknown species and impressive individuals. Of the 300,000 known species of plants, 60,000 grow as trees. Journeying to see new species and learning more about them makes for a lifetime of wonder and enjoyment.

I am a Californian. I was born here, as were my parents, grandparents, and several generations before them. Though I am fascinated by all plants, the native trees of this state are special for me, and always have been. They are numerous, varied, and fascinating. They define this place I call home. The purpose of this book is to share my love of trees and help guide the curious on their own journey of discovery.

California provides an opportunity, presented by few other places in the world, to observe impressive trees. Our species are superlatives: the most massive tree (giant sequoia), the longest-lived tree (Great Basin bristlecone pine), the world's largest pine (sugar pine), the tallest tree (redwood), the largest oak (canyon live oak), and the list goes on. In addition, more than one quarter of the tree species in the state only occur here!

Quercus lobata at Milpitas Special Interest Area, Los Padres N. F., Monterey County

A tree is not easily defined. It is a collection of billions of branches, root tips, buds, leaves, and layers of wood, all carrying out countless interactions with fungi, bacteria, insects, birds, mammals, and other plants. Trees perform the most complicated chemistry known by making intricate, high-energy molecules from the simple compounds of water and carbon dioxide. All of the interactions, relationships, and interdependencies inside and around a tree are not, nor may never be, fully understood.

There are also plenty of species (e.g., ceanothus and manzanitas) that occasionally satisfy any definition of a tree. I have omitted them, along with many of California's iconic large shrubs including toyon (*Heteromeles arbutifolia*), hollyleaf cherry (*Prunus ilicifolia*), smoke tree (*Psorothamnus spinosus*), elderberry (*Sambucus mexicana*). Nonetheless, they should rightly be revered as impressive plants. None of them care what their form is, and none have read this or any other book. They just are. In our feeble attempts to make sense of the world, we categorize, we make lists, we devise names, we are endlessly clever, and yet we always insufficiently describe what life actually is.

California is a large state, the eastern quarter of which is covered with deserts where trees rarely occur outside oases and watercourses. The remainder of California is forested, mostly by conifers in higher elevations or wetter areas, or blanketed with oak woodlands in the warmer and drier lower elevations. The remaining trees occur in coastal coniferous forests, in the coastal fog belt, or along rivers and streams.

Of California's 95 native tree species, 23 occur only in the state. Fifty-two species are conifers, 24 are deciduous hardwoods, 15 are evergreen, two are monocots (Joshua tree and California fan palm), and one is a cactus (saguaro). We have 18 of the country's 37 pines and 21 oak species. Ten of California's tree species are the only members of their genus. It is worth noting that less than two percent of California's approximately 5,000 native plants mature into trees.

The forests of California, like many similar ones worldwide, have been denigrated and destroyed by the graceless insensitivity and greed so common in our species. Our state's official tree, the ancient and towering redwood, was logged over more than 95% of its original range. Though these and other forests remain, they are in a much-diminished state. As I write, I hear daily news of the smoke that billows from our forests. With nearly 40 million people living in a hotter, drier California, and record wildfires raging in the wildlands every summer, California's native trees need public attention, intense management, and widespread protection now more than ever.

The purpose of this book is to promote greater knowledge and enjoyment of our native trees. California is changing. It is inevitable. Forests and trees are being lost to drought and fire at an unprecedented rate. Like a guide to the art in a museum that is burning, I hope this book helps us appreciate what we have while we still have it.

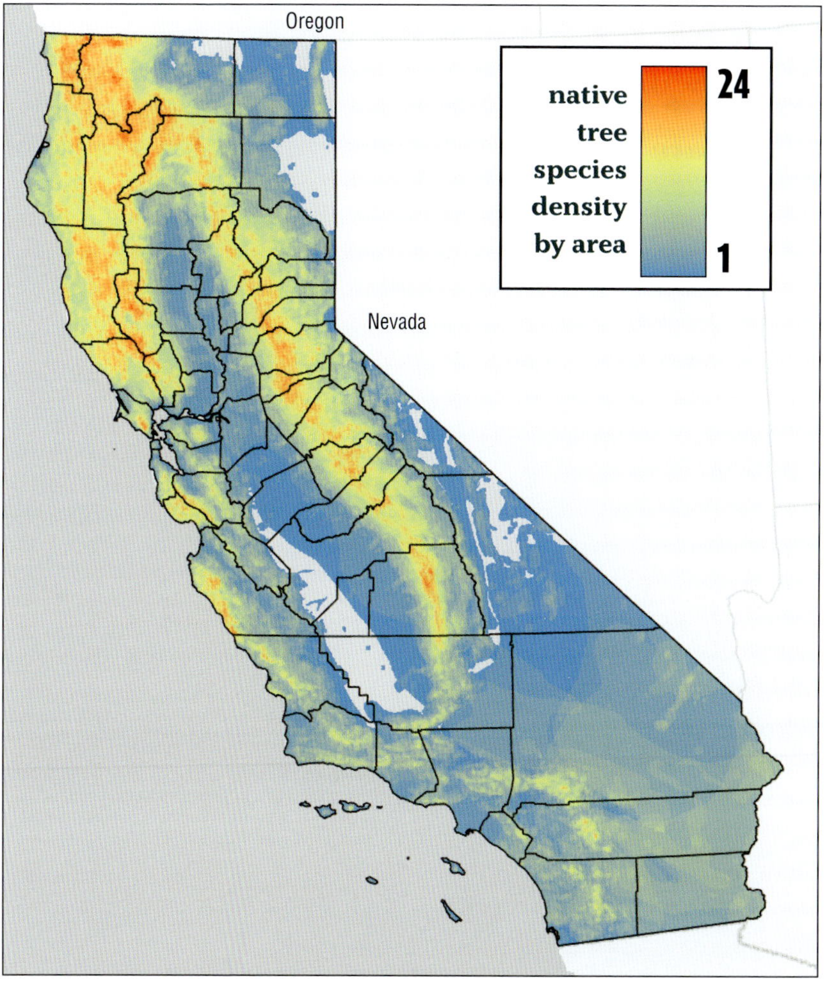

A heat map representing the number of native tree species across California. All 95 species are represented. Areas with warmer tones have higher numbers, while cooler tones have fewer. Areas with no native trees are light gray.

The highest density of native tree species in California occurs in northern California, in Siskiyou County, where the ranges of 24 species overlap. The highest density of endemic native tree species (those that only occur in California) is in Robert Louis Stevenson State Park, north of Santa Rosa, with eight species.

What Is a Tree?

"Trees are plants that people call trees—a term of dignity, not botany."

—Jared Farmer

We can define a tree in many ways. A tree is a plant you can walk under, while a shrub is a plant you must walk around. A tree is a plant that would hurt you if it fell on you. Tree, as used by Jared Farmer above, is a term of dignity, not botany.

For this book, we define a tree as a woody plant with a well-defined, solitary trunk and a raised canopy. For a large, woody plant species to be described as a tree here, over 90% of the total known or surveyed individuals in the species must mature to a height of over 20 feet and have a single, dominant trunk more than 6 inches in diameter at 4.5 feet above the ground.

Shrubs are smaller and shorter than trees and often have many small stems arising from near the ground. Each species treated as a tree in this book satisfies the definition of a tree most of the time. Some individuals of every tree species, due to environmental conditions, can grow as shrubs. Examples are windswept redwoods on the jagged mountains overlooking the Big Sur coast or high-elevation, knee-high mountain hemlocks bent under a century of snowpack.

Many shrubby species have some individuals that occasionally satisfy the definition of a tree. We have omitted those species from the list of 95 native trees of California, but by no means are these unimpressive plants. It is no surprise in a state with large areas of chaparral and coastal and desert shrublands that some of California's most iconic plants are shrubs. We treat those at the end of the book on page 182.

California's landscapes are often a mix of trees and shrubs, Pinnacles N. P.

California's Forests & Woodlands

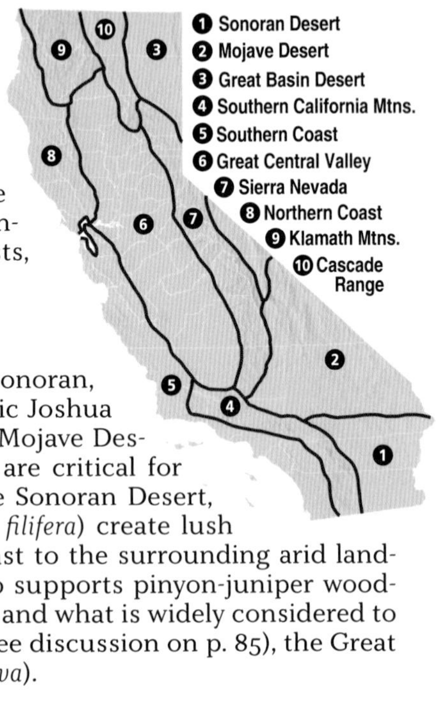

California's forests (mostly trees) and woodlands (trees and shrubs) are among the most diverse in the world, spanning ranges of climate, geology, elevation, and latitudes. From towering redwoods to southern palms, the state's native trees thrive in environments as varied as lush rainforests, dry deserts, and alpine peaks.

Desert Woodlands

California's deserts include the Sonoran, Mojave, and Great Basin. The iconic Joshua tree (*Yucca brevifolia*), found in the Mojave Desert, forms sparse woodlands that are critical for desert biodiversity. In oases of the Sonoran Desert, California fan palms (*Washingtonia filifera*) create lush habitats that stand in stark contrast to the surrounding arid landscape. The Great Basin Desert also supports pinyon-juniper woodlands and at the highest elevations and what is widely considered to be the Earth's oldest living thing (see discussion on p. 85), the Great Basin bristlecone pine (*Pinus longaeva*).

Southern California Forests

Southern California's forests are shaped by fire within a Mediterranean-type climate with hot, dry summers and mild, wet winters. The region's mountains, such as the San Bernardino and San Jacinto ranges, support mixed-conifer forests similar to those found in northern California. Lower elevations feature oak woodlands dominated by coast live oak (*Quercus agrifolia*) and valley oak (*Quercus lobata*). Chaparral ecosystems, characterized by drought-resistant shrubs and small trees, are also prominent in Southern California. The rare Tecate cypress (*Hesperocyparis forbesii*) represents a unique element of the region's chaparral diversity.

Northern California Forests

Northern California is home to some of the most emblematic forests in the state. Coastal regions feature the majestic redwoods (*Sequoia sempervirens*), the tallest trees on Earth, which thrive in the cool, moist climate provided by coastal fog. Inland areas, particularly in the Sierra Nevada and Klamath Mountains, are dominated by mixed-conifer forests, which include ponderosa pine (*Pinus ponderosa*), sugar pine (*Pinus lambertiana*), and Douglas-fir (*Pseudotsuga menziesii*).

California's Arboreal Legacy

The incredible diversity of California's forests and native trees reflects the state's unique ecological heritage. From the lush, towering redwoods of the north to the hardy Joshua trees of the desert, these forests are not only critical to the state's biodiversity but also hold cultural, historical, and ecological significance. What follows is a pictorial introduction to the tree diversity in each of these landscapes.

Mixed conifers along the South Yuba River, Nevada County

Sonoran Desert

The Sonoran Desert, located in southeastern California, is an arid region characterized by a hot, subtropical climate with mild winters (lack of freezing temperatures) and extremely high summer temperatures, often exceeding 110°F. Elevations range from below sea level to about 3,000 feet, influencing temperature and precipitation patterns. This desert is renowned for its diverse plant life, including trees such as the blue palo verde (*Parkinsonia florida*), which provides essential shade and habitat, and the ironwood (*Olneya tesota*), a keystone species supporting numerous animals and plants. Mesquite species are also prominent, thriving in washes and lowlands where water availability is higher. These hardy trees have adapted to the arid conditions in the desert through deep root systems and water-storing tissues.

7% of California

9 Native Trees

Olneya tesota in a wash, Whipple Mountains, San Bernardino County

Parkinsonia florida in a wash, San Bernardino County

Washingtonia filifera is restricted to the Sonoran Desert, San Bernardino County

Neltuma odorata in the Turtle Mountains, San Bernardino County

Mojave Desert

15% of California

17 Native Trees

The Mojave Desert, located primarily in southeastern California, is characterized by a harsh, arid climate with annual rainfall averaging between 3 and 5 inches, making it one of the driest deserts in North America. This limited precipitation falls mainly during the winter and early spring months, often in the form of brief, intense rainstorms. The timing of the rains and freezing temperatures are particularly challenging for plants, as they endure long periods without rainfall which, when it does occur, is in winter when plants are dormant. There are occasional monsoonal rains in summer.

The Mojave Desert is home to tree species that are specially adapted to survive these extreme conditions. The most emblematic tree is the Joshua tree (*Yucca brevifolia*) with its long, spiky leaves and twisted branches. Pinyon pines (*Pinus* sp.) are found at higher elevations or in slightly more favorable microclimates where moisture is more abundant. These trees, along with other drought-tolerant plants, have evolved a variety of strategies to conserve water, such as deep root systems, waxy leaf coatings, and the ability to go dormant during dry periods. Despite the challenges posed by limited rainfall, the Mojave Desert supports an amazing amount of plant diversity. In fact, if the Mojave had political boundaries that made it a state, it would be the 8th most diverse state in the U.S. in terms of plant species.

The New York Mountains hold remnant woodlands similar to coastal So. California

Pinus monophylla in the Mojave National Preserve, San Bernardino County

Yucca jaegeriana woodland in the Mojave National Preserve, San Bernardino County

Great Basin

12% of California

16 Native Trees

The Great Basin Desert in California is characterized by its arid climate, rugged terrain, and distinct plant communities. Stretching into parts of Modoc, Lassen, Mono, and Inyo counties, this represents the western edge of the larger Great Basin—which spans much of Nevada and Utah while reaching into parts of Oregon and Idaho. The California portion endures long periods of freezing temperatures and is in the rain shadow of the Cascades and Sierra. It encompasses high desert plateaus, sagebrush steppes, and scattered mountain ranges, shaped by variations in elevation and geology.

In the north, native tree species such as the western juniper (*Juniperus occidentalis*) dominate the landscape, particularly in rocky uplands. Drought-tolerant trees are well-adapted to the high desert, providing habitat for birds, insects, and small mammals. Ponderosa pine (*Pinus ponderosa*), lodgepole pine (*Pinus contorta*), and whitebark pine (*Pinus albicaulis*) are present at higher elevations where moisture levels are slightly greater. In the lower elevations, singleleaf pinyon pine (*Pinus monophylla*) and Utah juniper (*Juniperus osteosperma*) form vast pinyon-juniper woodlands. In higher elevations in the White and Inyo mountains limber pine (*Pinus flexilis*) and Great Basin bristlecone pines (*Pinus longaeva*) are found.

Pinus albicaulis high in the Warner Mountains, Modoc County

High elevations hold limber and Great Basin Bristlecone pines, White Mtns., Inyo Co.

Juniperus osteosperma near the Nevada state line, Modoc County

Southern California Mtns.

8% of California

36 Native Trees

The southern California Mountains are composed of a number of distinct ranges. The Transverse Ranges, which include the San Bernardino, San Gabriel, and Santa Ynez Mountains, are unusual in that they are orientated east-west—a dramatic contrast to the generally north-south orientation of most of California's mountains. This unique alignment is the result of tectonic forces, where the Pacific and North American plates have pushed the land in different directions. The climate of these mountains, as well as the north-south-trending Peninsular Ranges east of San Diego, is variable due to elevation gradients—hotter, drier conditions below and cooler, wetter conditions higher up.

Ecologically, these mountains support a diverse mix of habitats, from chaparral and coastal sage scrub at lower elevations to coniferous forests at higher altitudes. Tree species are highly variable, with quintessential species like coast live oak (*Quercus agrifolia*) and southern California black walnut (*Juglans californica*) in the lower foothills and towering pines such as Jeffrey pine (*Pinus jeffreyi*) and sugar pine (*Pinus lambertiana*) higher up. At mid-to-higher elevations, the endemic bigcone Douglas-fir (*Pseudotsuga macrocarpa*) is a notable species, uniquely adapted to the region. Other tree species, such as incense-cedar (*Calocedrus decurrens*) and black oak (*Quercus kelloggii*), can be found in cooler, more moderate zones.

Pinus contorta in the San Gabriel Mountains, Los Angeles County

Pinus ponderosa, Pseudotsuga macrocarpa, and *Quercus kelloggii* in the Sierra Pelona Mountains above the Mojave Desert, Los Angeles County

Pinus ponderosa in the Laguna Mountains above the Sonoran Desert, San Diego Co.

Pinus coulteri, Santa Ynez Mountains, Santa Barbara County

Southern Coast

The Southern Coast Ranges, defined here from the San Francisco Bay Area to San Diego, represent a diverse and ecologically significant region. They are part of the larger Coast Ranges of California, a geologically active area shaped by the tectonic forces of the Pacific and North American plates. Geology and climate shape varied landscapes that range from grasslands to dense woodlands and forests.

Ecologically, the Southern Coast Ranges are home to a variety of native tree communities. The most prominent native trees in the area include coast live oak (*Quercus agrifolia*), valley oak (*Quercus lobata*), and Shreve oak (*Quercus parvula* var. *shrevei*). The area is home to four tree species that grow nowhere else: Monterey pine (*Pinus radiata*), Monterey cypress (*Hesperocyparis macrocarpa*), Gowen cypress (*Hesperocyparis goveniana*), and Santa Lucia fir (*Abies bracteata*).

Abies bracteata and *Pinus lambertiana* in the Santa Lucia Mountains, Monterey Co.

Pinus radiata and *Hesperocyparis macrocarpa* at Point Lobos S. N. R., Monterey Co.

Pinus torreyana on Santa Rosa Island, Santa Barbara County

Pinus ponderosa and *Pinus sabiniana* in a mixed evergreen woodland, Monterey Co.

Great Central Valley and Foothills

15% of California

16 Native Trees

California's Great Central Valley and its surrounding foothills represent a vast and geologically diverse region that plays a critical role in the state's ecology, agriculture, and cultural history. Stretching over 400 miles from the northern Sacramento Valley to the southern San Joaquin Valley, this area is bordered by the Sierra Nevada to the east and the Coast Ranges to the west. The valley is one of the most fertile agricultural regions in the world, underpinned by alluvial soils deposited over millennia by rivers such as the Sacramento, San Joaquin, and their tributaries. The valley floor averages only about 20 inches of rain annually in the north, while the southern sections, particularly around Bakersfield, receive as little as 6 inches, placing parts of the valley in a semi-arid category.

Native trees of the region reflect its diverse habitats and climate. Valley oaks (*Quercus lobata*) are emblematic to the Great Central Valley. Blue oaks (*Quercus douglasii*) and interior live oaks (*Quercus wislizeni*) are prominent in the foothills, adapted to the drier, rockier soils. California sycamores (*Platanus racemosa*) and Fremont cottonwoods (*Populus fremontii*) are common near waterways, providing critical habitat for wildlife. In higher elevations of the foothills, ghost pines (*Pinus sabiniana*), with their sparse, spindly appearance, reflect the region's adaptation to periodic drought and fire.

Blue oak woodland in eastern San Luis Obispo County

Remnant valley oak stand in Monterey County

Riparian trees and oak woodlands in fall, Southern Sierra Nevada foothills, Kern Co.

Sierra Nevada

12% of California

41 Native Trees

The Sierra Nevadais renowned for its striking landscapes, ecological diversity, and geological complexity. Spanning approximately 400 miles from north to south, this range includes some of the most iconic natural landmarks in the United States, such as Yosemite Valley, Lake Tahoe, and Mount Whitney. Its ecology, climate, and geology create distinct vegetation communities, especially among native trees, that define the region's character.

Above the foothills, tree communities transition into mixed-conifer forests. Common species include ponderosa pine (*Pinus ponderosa*), sugar pine (*Pinus lambertiana*), Douglas-fir (*Pseudotsuga menziesii*), white fir (*Abies concolor*), and red fir (*Abies magnifica*). One of the most exemplary trees in the Sierra Nevada is the giant sequoia (*Sequoiadendron giganteum*), which grows in limited groves on the western slopes of the range between 4,000 and 8,000 feet. Notable groves include the Mariposa Grove in Yosemite National Park and the Giant Forest in Sequoia National Park. Higher up, subalpine forests take over, dominated by lodgepole pine (*Pinus contorta*), whitebark pine (*Pinus albicaulis*), foxtail pine (*Pinus balfouriana*), and mountain hemlock (*Tsuga mertensiana*).

Pinus balfouriana in the subalpine of Sequoia National Park, Tulare County

Fall in a *Sequoiadendron giganteum* grove, Sequoia N. P. Tulare County

Sierra Nevada mixed-coniferous forest, Tahoe National Forest, Nevada County

Conifers ringing Emerald Bay, Lake Tahoe, El Dorado County

Northern Coast

8% of California

46 Native Trees

California's Northern Coast Ranges stretch from the San Francisco Bay Area to the Oregon border, forming a rugged and ecologically rich region. These coastal mountains are part of the larger Coast Ranges and represent a transition zone between California's Mediterranean ecosystems and the lush temperate rainforests of the Pacific Northwest. The area's dramatic landscapes include rocky shorelines, dense forests, and rolling hills—shaped by tectonic activity and proximity to the Pacific Ocean.

This region is home to some of the most emblematic forests in North America, including the southern extent of the Pacific Northwest temperate rainforest. Towering redwoods (*Sequoia sempervirens*) dominate much of the coastal landscape. Other notable tree species include Sitka spruce (*Picea sitchensis*), Douglas-fir (*Pseudotsuga menziesii*), and western hemlock (*Tsuga heterophylla*), which are more common in the northern reaches closer to Oregon. Oak woodlands and mixed-evergreen forests also occur inland, featuring tanoak (*Notholithocarpus densiflorus*), Pacific madrone (*Arbutus menziesii*), and bay laurel (*Umbellularia californica*).

Eel River headwaters, Yolla Bolly-Middle Eel Wilderness, Tehama County

Prairie Creek Redwoods National and State Park, Humboldt County

Lower Eel River, Humboldt Redwoods State Park, Humboldt County

Klamath Mountains

7% of California

46 Native Trees

The Klamath Mountains, located in northern California and extending into southern Oregon, are a rugged and ecologically rich region renowned for hosting one of the most diverse temperate coniferous forests on Earth. This mountainous landscape, shaped by millions of years of geological activity and climatic variation, supports an incredible array of plant and animal life.

The Klamath Mountains boast a staggering diversity of conifer species. Iconic trees like Douglas-fir (*Pseudotsuga menziesii*), sugar pine (*Pinus lambertiana*), and incense-cedar (*Calocedrus decurrens*) dominate the mixed-conifer forests, while endemic species such as Brewer spruce (*Picea breweriana*) and Port Orford-cedar (*Chamaecyparis lawsoniana*) add to the region's uniqueness. At lower elevations, hardwoods like tanoak (*Notholithocarpus densiflorus*), Pacific madrone (*Arbutus menziesii*), and canyon live oak (*Quercus chrysolepis*) mix with conifers to create diverse forests. Higher elevations support white fir (*Abies concolor*), Shasta red fir (*Abies magnifica* var. *shastensis*), and mountain hemlock (*Tsuga mertensiana*).

Smith River headwaters, Six Rivers National Forest, Del Norte County

Marble Mountain Wilderness, Salmon River headwaters, Siskiyou County

Mixed-evergreen lowland forest along the Trinity River, Trinity County

Russian Wilderness, Scott River headwaters, Siskiyou County

Cascade Range

4% of California

25 Native Trees

The Cascade Range, located in north-central California and extending into Oregon and Washington, is part of the larger Cascade volcanic arc. This region is defined by its volcanic origins, with emblematic peaks like Mount Shasta and Lassen Peak dominating the landscape. These mountains are surrounded by diverse forests and ecosystems, influenced by their elevation, geology, and proximity to both the Pacific Ocean and the arid Great Basin.

The forests of California's Cascade Range are characterized by a mix of coniferous species adapted to the varied conditions of the region. At lower elevations, ponderosa pine (*Pinus ponderosa*), Douglas-fir (*Pseudotsuga menziesii*), and incense-cedar (*Calocedrus decurrens*) dominate the landscape. Mid-elevation forests transition into mixed-conifer communities that include white fir (*Abies concolor*) and sugar pine (*Pinus lambertiana*). Higher elevations support subalpine forests with species like mountain hemlock (*Tsuga mertensiana*), lodgepole pine (*Pinus contorta*), whitebark pine (*Pinus albicaulis*), and western white pine (*Pinus monticola*).

Pinus albicaulis and *Pinus contorta* on the Goosenest, Siskiyou County

Lassen Peak, Lassen National Park, Tehama County

Pinus albicaulis on a volcano north of Mount Shasta near Oregon, Siskiyou County

Meadow complex north of Mount Shasta rimmed with conifers, Siskiyou County

Coniferous Trees

Conifers are a group of ancient, mostly evergreen trees and shrubs, known for their cone-bearing reproductive structures and needle-like or scale-like leaves. They belong to plant families that include pines, firs, cedars, spruces, and redwoods. Conifers have been on Earth since the time of the dinosaurs and are thus well adapted to diverse environments across the globe. They represent less than 1% of all plant species yet still cover 40% of the forested land on Earth! In California, conifers play a crucial role in shaping the state's diverse landscapes. From the misty coastal ranges that are home to towering coast redwoods, to the foxtail pines found in the mountainous Sierra Nevada, and the ancient bristlecone pines that stand resilient at the edge of the Great Basin, conifers thrive across a variety of geological formations and elevations. Their impressive adaptability is a testament to their long evolutionary history. In fact, California nurtures some of the highest conifer species diversity on Earth.

California's conifers are not just visually striking, they are integral to the state's ecological balance. They are capable of surviving extreme conditions, with some species adapted to coastal fog, others to high altitudes, and many to the state's seasonal droughts and fire cycles. Conifers are both emblematic of natural beauty and indispensable to ecological health and are deeply rooted in California's history, culture, and identity. Two of California's nine national parks are named after conifers.

In California, conifers are represented by three families: **Cupressaceae, Pinaceae,** and **Taxaceae**.

Cypress Family

The Cupressaceae includes trees with both scale-like and needle-like leaves and cones ranging from small and woody to fleshy and berry-like. Some of California's most famous trees belong to this family, including the redwood, the world's tallest tree, which thrives in the fog-laden coastal forests from central California northward. The family also includes the giant sequoia, which is notable for being the world's largest living plant. Other members of this family in California include the false cedars, cypresses, and junipers.

The Cupressaceae is the most widely distributed conifer family in the world, occurring in all major habitats in both the Southern and Northern Hemispheres. It is the largest family in terms of genera and third largest in terms of species. Monotypic genera (a genus with a single species) typify the family. Many species have become isolated over time in disjunct populations, becoming the sole survivors of lineages that once had many more species.

The major exception to this pattern is in the junipers (*Juniperus* spp.), the largest genus in the family. Because of a mutualistic relationship with birds, which disperse their seeds and help overcome barriers otherwise imposed by water or wind, the 50 species of juniper have a nearly continuous distribution across the mid-latitudes of the Northern Hemisphere. While no members of the Cupressaceae attain dominance over immense geographic ranges, as do the spruces or firs within Pinaceae, they can become dominant on a local scale.

False Cedars (p. 40-44)

Cypresses (p. 45-55)

Junipers (p. 56-60)

Redwood and Sequoia (p. 61-65)

False Cedars

California has four species in a group of trees sometimes called false cedars: yellow-cedar (*Callitropsis nootkatensis*), incense-cedar (*Calocedrus decurrens*), Port Orford-cedar (*Chamaecyparis lawsoniana*), and western redcedar (*Thuja plicata*). Forty different trees around the world bear the meaningless common name cedar for having aromatic, spicy-scented wood or a superficial resemblance to the four species of true cedar in the genus *Cedrus* (Pinaceae).

Calocedrus decurrens and *Chamaecyparis lawsoniana* grow together on serpentine soils in the Siskiyou Mountains, Siskiyou County

YELLOW-CEDAR

Callitropsis nootkatensis

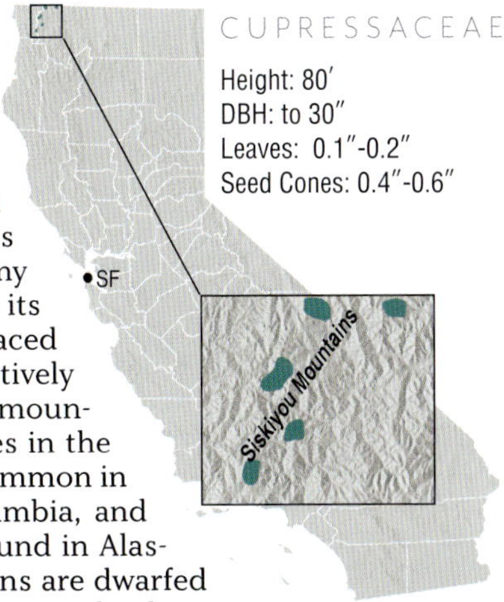

CUPRESSACEAE

Height: 80′
DBH: to 30″
Leaves: 0.1″-0.2″
Seed Cones: 0.4″-0.6″

Yellow-cedar, also called Nootka cypress, has scale-like leaves in pairs on flat, drooping branches and seed cones the size of a small grape. The taxonomy of this species has been debated since its discovery, and it has been variously placed in different genera. Yellow-cedar is relatively rare in California, occurring only in the mountains of Del Norte and Siskiyou counties in the Siskiyou Mountains. It is much more common in Oregon, Washington, and British Columbia, and ranges northward to Prince William Sound in Alaska. While enormous up north, specimens are dwarfed in California. It has a tolerance for heavy snow loads.

ID Tips: crown narrow, cone-shaped, branches drooping; bark thin, grayish-brown; needles small, scale-like; in CA only in Siskiyou Mountains

INCENSE-CEDAR

Calocedrus decurrens

CUPRESSACEAE

Height: 100′-130′
DBH: 30″-70″
Leaves: 0.01″-0.08″
Seed Cones: 0.75″-1.4″

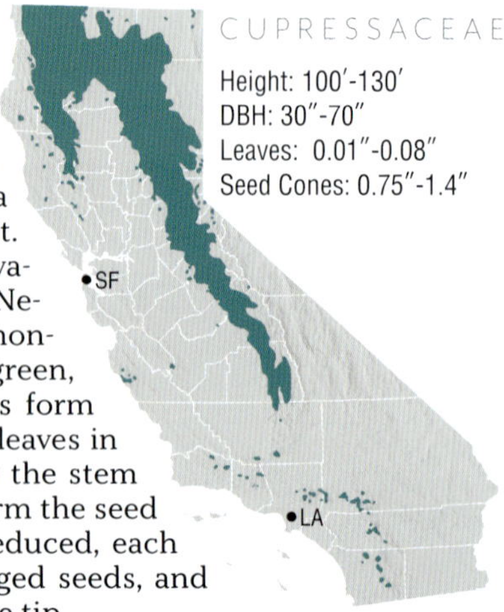

This species grows in mixed-coniferous forests from northern Baja California to the southern slopes of Oregon's Mt. Hood. It is one of the dominant mid-elevation, western-slope trees of the Sierra Nevada. It has beautifully tapering, cinnamon-red trunks and a crown of yellowish-green, drooping foliage. Clusters of branchlets form flattened sprays with minute, scale-like leaves in two pairs at each stem node that cover the stem completely. Only three pairs of scales form the seed cones: the outermost pair are highly reduced, each scale of the second pair bears two winged seeds, and the two innermost scales are fused to the tip.

ID Tips: bark cinnamon-red, forming ridges; leaves small, scale-like, branchlets in flattened sprays; seed cones like duck beaks

PORT ORFORD-CEDAR

Chamaecyparis lawsoniana

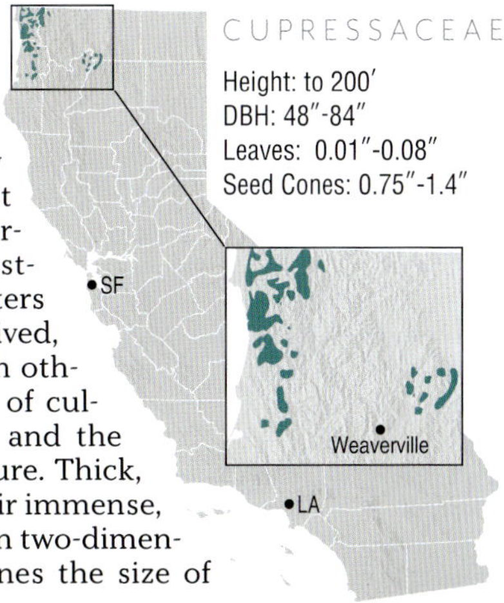

CUPRESSACEAE

Height: to 200′
DBH: 48″-84″
Leaves: 0.01″-0.08″
Seed Cones: 0.75″-1.4″

Port Orford-cedars occur primarily along a 200-mile run of the Pacific coast from Humboldt County to Coos Bay, Oregon, with inland populations in the eastern Klamath Mountains in the headwaters of the Sacramento River. They are long-lived, massive trees, and mostly co-occur with other conifers in mixed forests. Hundreds of cultivated varieties are grown in Europe, and the heartwood is prized in Japan for furniture. Thick, soft straps of light brown bark cover their immense, tapering trunks. The lacy foliage grows in two-dimensional sprays with bluish-gray seed cones the size of small grapes.

ID Tips: leaf undersides with stomatal bloom forming X shape; seed cones spherical, bluish-green; foliage smells ginger-like when crushed

WESTERN REDCEDAR

Thuja plicata

CUPRESSACEAE

Height: to 200′
DBH: to 180″
Leaves: 0.01″-0.08″
Seed Cones: 0.5″-0.9″

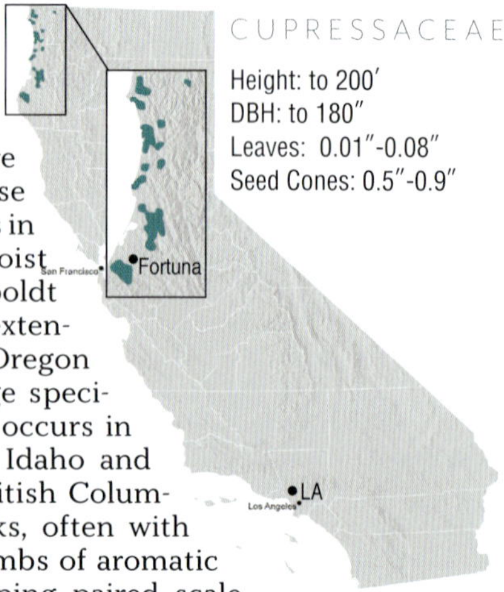

Of the six species of *Thuja*, only two are found in North America, and one of these occurs in California. The largest species in its genus, western redcedar grows on moist slopes and in deep gulches from Humboldt County northward to Alaska, forming extensive stands of massive trees along the Oregon and Washington coasts. However, large specimens are rare in California. Inland, it occurs in the northern Rockies from northern Idaho and western Montana into Alberta and British Columbia. Tapered, cinnamon-colored trunks, often with buttressed bases, support drooping limbs of aromatic foliage. Branchlets with tightly overlapping, paired, scale-like leaves grow in flat sprays. Small, egg-shaped, few-seeded cones cluster upright on the branchlets.

ID Tips: bark fissured into vertical bands; leaf undersides with stomatal bloom creating butterfly shape; seed cones upright, yellow-green

Cypresses

Hesperocyparis is the genus for 16 species native to western North America, all of which were formerly classified in the genus *Cupressus*. Ten of these species are native to California, and eight are restricted almost entirely to the state. Many are rare, occurring only in isolated populations on poor soils. Though their ranges rarely overlap, they can still be difficult to distinguish, and experts disagree on the total number of species present in California. Cypresses are closed-cone conifers whose cone scales are sealed with pitch until opened by the heat of a fire—a strategy allowing for germination and regrowth in the absence of competition from established adult trees. The bark is either fibrous with brown to gray strips or smooth and peeling in thin flakes. Their small, aromatic leaves are arranged in opposite pairs. Woody, brownish seed cones and smaller, yellow pollen cones form on branch tips of the same tree. Several of these species have sen stand replacing fires in the 21st century. But, even in these places, regeneration is abundant due to species' adaptation to, and dependence on, events like these.

Hesperocyparis macrocarpa at Point Lobos S. N. R., Monterey County

SANTA CRUZ CYPRESS

Hesperocyparis abramsiana

CUPRESSACEAE

Height: to 75′
DBH: 6″-12″
Leaves: to 0.5″
Seed Cones: 0.75″-1.2″

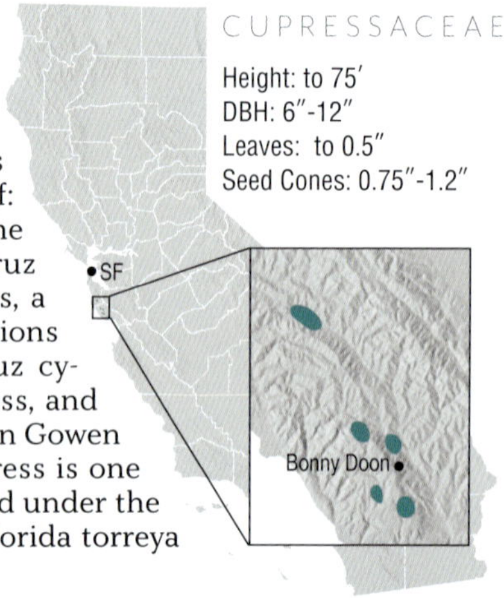

When trying to identify wild cypresses in California, you must first ask yourself: where am I standing? If you are on the foggy western slopes of the Santa Cruz Mountains, it is the Santa Cruz cypress, a species that occurs in five small populations on sandy soils in chaparral. Santa Cruz cypress is closely related to Gowen cypress, and some experts think it is a hybrid between Gowen and Sargent cypresses. Santa Cruz cypress is one of only two native U.S. conifers protected under the Endangered Species Act (the other is Florida torreya [*Torreya taxifolia*] from Florida).

ID Tips: bark fibrous, grayish, forming vertical strips; leaves small, scale-like, bright green, fragrant; Santa Cruz Mountains

BAKER CYPRESS

Hesperocyparis bakeri

CUPRESSACEAE

Height: 30′- 82′
DBH: to 20″
Leaves: 0.1″-0.2″
Seed Cones: 0.4″-1″

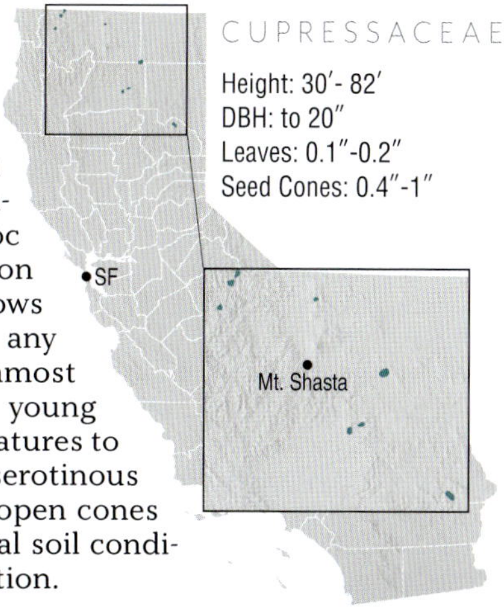

The Baker cypress occurs in California and southwestern Oregon in the Klamath Mountains, Cascade Range, Modoc Plateau, and northern Sierra Nevada on volcanic rock and serpentine soils. It grows at the highest elevations reached by any California cypress and is the northernmost species in the genus. The new bark of young trees is bright orange and peeling but matures to a thick, brown, alligator-like skin. It has serotinous cones, and relies upon frequent fire to open cones and to create the high-light, bare mineral soil conditions that the seed requires for germination.

ID Tips: bark thin, red when young, gray in maturity; foliage of sparse, blue-green sprays, fragrant

TECATE CYPRESS

Hesperocyparis forbesii

CUPRESSACEAE

Height: to 30′
DBH: to 18″
Leaves: to 0.05″
Seed Cones: 0.8″-1.2″

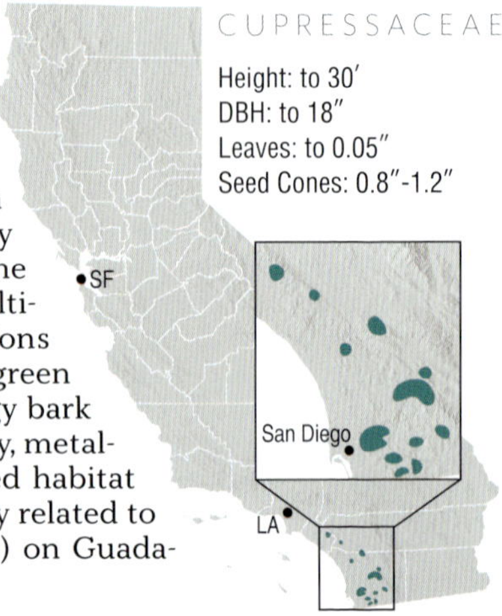

Tecate cypress is primarily a Baja California species, and most populations in California occur on the slopes of Otay Mountain, southeast of San Diego near the Mexican border. It can grow as a large, multibranched shrub but in favorable conditions reaches up to 30 feet in height. Its dusty green foliage grows on trunks with thin, shaggy bark that sheds to expose smooth orange, gray, metallic green, and red patches. The preferred habitat is chaparral from 750 to 4000 feet. Closely related to the Guadalupe cypress (*H. guadalupensis*) on Guadalupe Island, Mexico.

ID Tips: bark exfoliating from trunk; leaves dusty green (never glaucous); pollen cones abundant; spherical seed cones brown-gray

GOWEN CYPRESS

Hesperocyparis goveniana

CUPRESSACEAE

Height: generally < 40′
DBH: to 20″
Leaves: 0.08″-0.2″
Seed Cones: 0.4″-0.8″

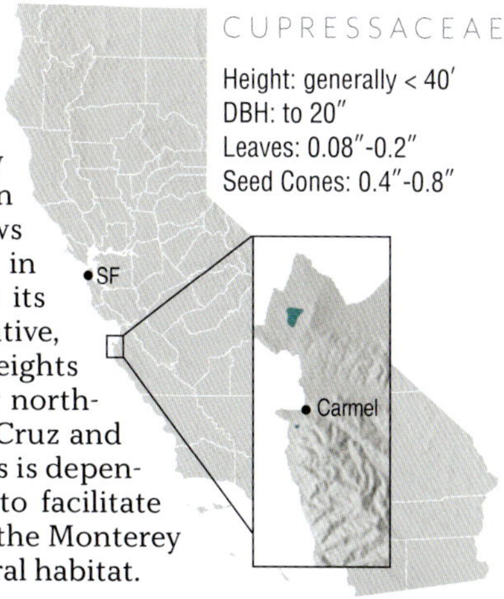

Gowen cypress has a limited range, only occurring in two small populations on the Monterey Peninsula where it grows among the coastal closed-cone pines in poorly drained, acidic soil that stunts its growth. Typically this species is diminutive, but on more fertile soils can grow to heights of 100 feet. Closely related to the other northern California coastal cypresses, Santa Cruz and pygmy. Like other cypresses, this species is dependent on frequent fire-return intervals to facilitate regeneration. Visit Huckleberry Hill on the Monterey Peninsula to see this species in its natural habitat.

ID Tips: bark reddish to grayish-brown with linear furrows; leaves light green to yellowish-green, fragrant; seed cones dark brown

McNAB CYPRESS

Hesperocyparis macnabiana

CUPRESSACEAE

Height: generally < 30′
DBH: to 12″
Leaves: 0.08″-0.2″
Seed Cones: 0.4″-0.8″

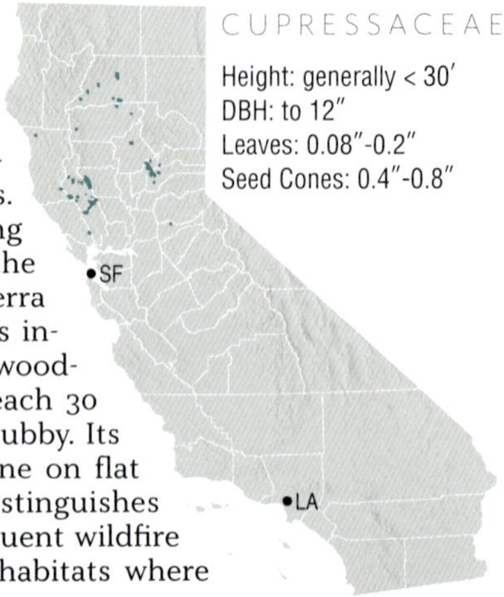

McNab cypress is one of the most widely occurring of the California cypresses. It is found in the foothills surrounding the northern Sacramento Valley to the Cascade Range and the northern Sierra Nevada. It generally grows in poor soils including serpentine in chaparral and woodlands. Although McNab cypress can reach 30 to 50 feet in height, it is frequently shrubby. Its grayish-green, resin-dotted foliage, borne on flat rather than three-dimensional sprays distinguishes it from other California cypresses. Frequent wildfire fire-return intervals are typical in the habitats where this species is found.

ID Tips: bark shredding, furrowed when mature; leaves grayish-green, with distinct resin dots, fragrant

CONIFEROUS TREES

MONTEREY CYPRESS

Hesperocyparis macrocarpa

CUPRESSACEAE

Height: to 130′
DBH: to 90″
Leaves: 0.08″-0.1″
Seed Cones: 0.8″-1.2″

This is among the most widely cultivated trees in the world, but is rare in the wild. It occurs naturally only on foggy, windswept slopes near Pebble Beach and Carmel in Monterey County. Fossil records show that surviving stands are remnants of more extensive forests that once grew along the west coast of North America. It has been moved around the world by humans to colonize coastal areas including most of the Pacific Coast of the U.S., New Zealand, and the Mediterranean region. Old cones are visible on the branches.

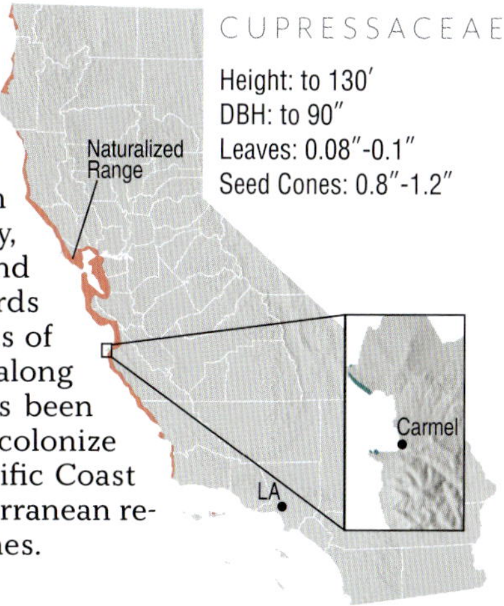

ID Tips: bark reddish-brown with shallow furrows; leaves scale-like, dark green, lemony fragrant; mature seed cones spherical, greenish-brown

PIUTE CYPRESS

Hesperocyparis nevadensis

CUPRESSACEAE

Height: to 30′
DBH: to 20″
Leaves: 0.08″-0.2″
Seed Cones: 0.4″-0.8″

Piute cypress is a long-lived species that only grows in several groves in the dry southern Sierra Nevada around Lake Isabella. Its gray, resinous foliage is highly aromatic when crushed. It is found on granitic soils from 3,000 to 6,000 feet in pinyon-juniper and oak-pine woodlands or chaparral. Some lump this species into the more widely distributed Arizona cypress (*H. arizonica*), which occurs in Arizona and Mexico.

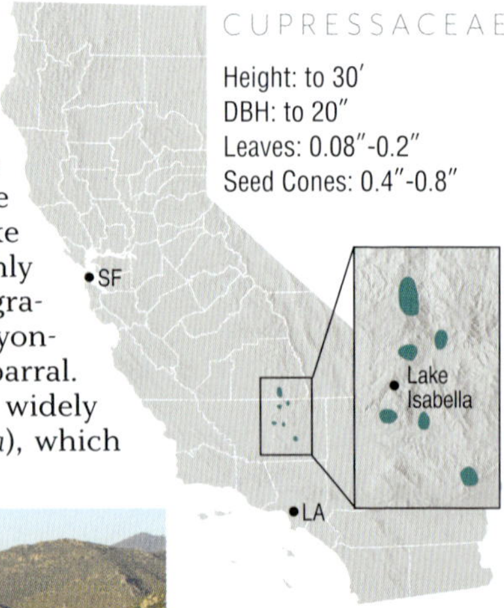

ID Tips: foliage sparse, fragrant, gray-green to blue-green; leaves highly resinous; southern Sierra Nevada

PYGMY CYPRESS

Hesperocyparis pigmaea

CUPRESSACEAE

Height: 10′-150′
DBH: to 12″
Leaves: 0.05″-0.1″
Seed Cones: 0.4″-0.8″

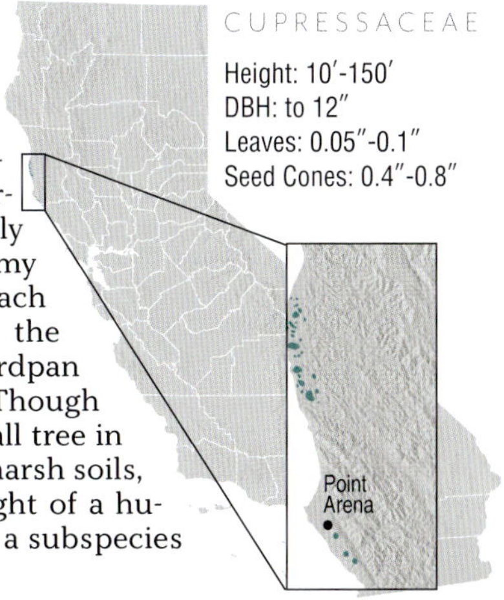

Some places have soils so harsh and devoid of nutrients that few plants can survive, and those that do become severely stunted. Such is the case for the pygmy cypress, which occurs on ancient beach terraces in Mendocino County. There the acidic, white-sandy soils occur over a hardpan of clay that prevents root penetration. Though the pygmy cypress can mature into a tall tree in favorable conditions, on these shallow, harsh soils, it takes many years to grow to the height of a human. Some experts treat this species as a subspecies of its close relative, Gowen cypress.

ID Tips: small, stunted size (usually); foliage scale-like, dark green; seed cones almost spherical; preference for acidic, nutrient-poor soils

SARGENT CYPRESS

Hesperocyparis sargentii

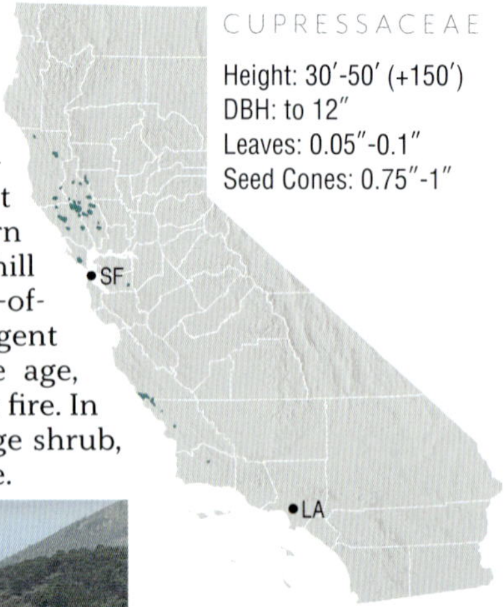

CUPRESSACEAE

Height: 30′-50′ (+150′)
DBH: to 12″
Leaves: 0.05″-0.1″
Seed Cones: 0.75″-1″

This is the most widely distributed cypress in California, growing in the Coast Ranges from Santa Barbara to northern Mendocino counties in chaparral, foothill woodland, and lower montane forests—often on serpentine. Many stands of Sargent cypress contain trees of all the same age, based on the time since the most recent fire. In dry sites this species may grow as a large shrub, but in moist soil it can grow as a tall tree.

ID Tips: often shrubby; bark furrowed; leaves small, scale-like; seed cones brown, remaining closed until opened by fire

CUYAMACA CYPRESS

Hesperocyparis stephensonii

CUPRESSACEAE

Height: 25′-45′
DBH: to 10″
Leaves: 0.05″-0.1″
Seed Cones: 0.75″-1″

In California, Cuyamaca cypress grows on a chaparral-covered slope in one small population in San Diego County, making it the tree with the most restricted native range of any in the state. Compared to its close relative the Tecate cypress, it has larger, more knobby cones, rougher foliage, and sheds its pollen at a different time. Its bark is thin, light gray, and sheds to expose reddish-brown patches.

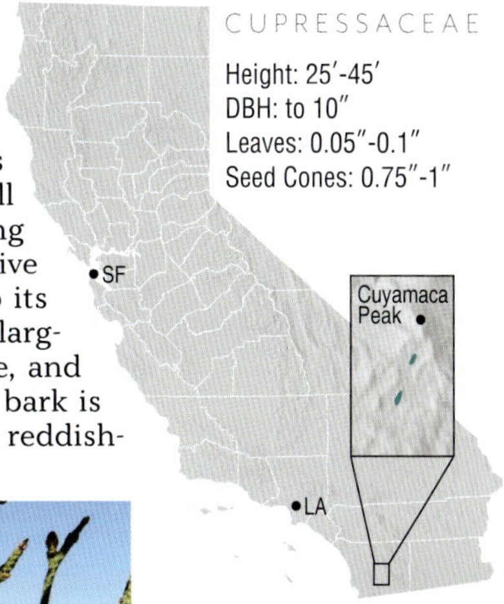

ID Tips: spreading tree with a central leader, only slightly taller than it is wide; only near Cuyamaca Peak, San Diego County

CONIFEROUS TREES

Junipers

The 50 species of juniper occur widely across the Northern Hemisphere. Five species occur in California, three of which (*J. grandis, J. occidentalis,* and *J. osteosperma*) regularly grow into trees. California juniper (*J. californica*) matures into a large shrub, rarely forming a single dominant trunk. The common juniper (*J. communis*) is a low, sprawling shrub with sharply pointed, needle-like leaves. Most junipers make tiny, scale-like leaves in pairs or rings of three. Birds eat and disperse the seeds inside the pea-sized, fleshy cones. *Juniperus communis* cones are used in flavoring gin. Some juniper species (e.g., *J. grandis*) have separate male and female individuals, with the fleshy female cones only appearing on half the trees, whereas others (e.g., *J. osteosperma*) have both male and female cones on the same tree.

Juniperus occidentalis in the Yolla Bolly-Middle Eel Wilderness, Tehama County

WESTERN JUNIPER

Juniperus occidentalis

CUPRESSACEAE

Height: 15′-50′ (+90′)
DBH: 10″-30″
Leaves: 0.1″-0.2″
Seed Cones: 0.3′-0.4″

Western juniper grows on the arid hills and plateaus of northwestern Nevada, eastern Oregon, western Idaho, and Washington. In California, it dominates the dry juniper woodlands in the northeastern part of the state with isolated populations in the Coast Ranges and Klamath Mountains. Western juniper is a slow-growing, long-lived, short-statured tree with shreddy bark, cones resembling blueberries, and bluish-gray foliage rough to the touch. Each scale-leaf has a sticky, white, aromatic resin dot on its surface. Western juniper is closely related to Sierra juniper (*J. grandis*) and both have been treated as varieties of this species.

ID Tips: bark reddish-brown, fibrous, shredding; leaves small, scale-like, gray-green; seed cones small, berry-like, blue-black

CONIFEROUS TREES

SIERRA JUNIPER

Juniperus grandis

CUPRESSACEAE

Height: 40′-80′
DBH: 12″-40″ (+120″)
Leaves: 0.05″-0.1″
Seed Cones: 0.1″-0.2″

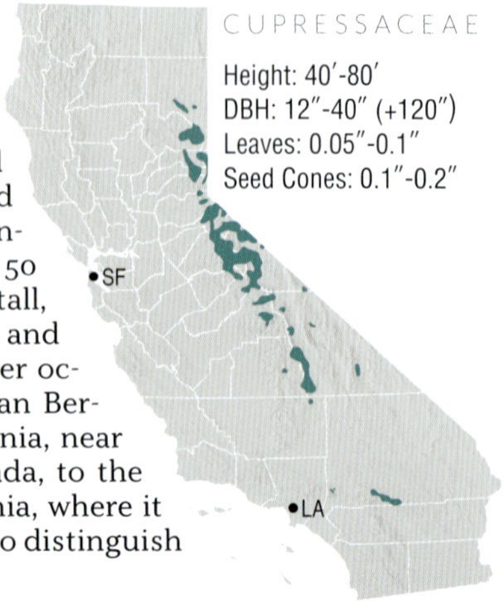

Sierra juniper is a striking, red-barked tree often growing twisted and wind pruned in the rocky fissures of open granite fields. It is the largest of the world's 50 junipers, and though not usually very tall, individuals can form enormous trunks and live for thousands of years. Sierra juniper occurs from the San Gabriel (rare) and San Bernardino Mountains of southern California, near Big Bear Lake, through the Sierra Nevada, to the Modoc Plateau in northeastern California, where it intergrades with and becomes difficult to distinguish from western juniper (*J. occidentalis*).

ID Tips: bark reddish-brown; leaves small, scale-like, dark green; seed cones berry-like, blue-black with a whitish bloom, maturing red-brown

The Bennett Juniper

This remarkable tree in the Stanislaus National Forest is the largest and oldest known Sierra juniper (*Juniperus grandis*) in the world, with an estimated age of over 2,000 years. Standing at over 80 feet tall, it has a massive, gnarled trunk, measuring about 12 feet in diameter. The tree is named after naturalist Clarence Bennett, who brought attention to its significance in the 1930s.

UTAH JUNIPER

Juniperus osteosperma

CUPRESSACEAE

Height: 10′-20′ (+35′)
DBH: 10″-20″
Leaves: 0.04″-0.08″
Seed Cones: 0.1′-0.2″

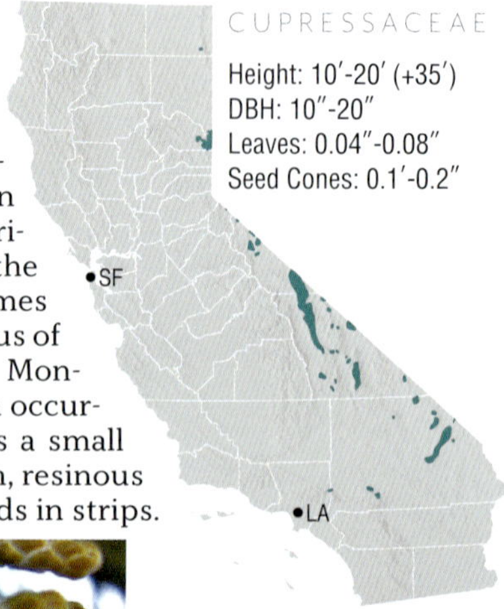

Utah juniper is one of the most widespread and hardy trees in the American west. In California, it has a spotty distribution in the eastern mountains of the Mojave and Great Basin deserts. It becomes more common on dry slopes and plateaus of the Great Basin, ranging northward to Montana and eastward to New Mexico, often occurring with pinyon pines. Utah juniper is a small tree with a rounded crown of olive green, resinous foliage and grayish-brown bark that sheds in strips.

ID Tips: trunk many-forked; crown generally bushy, rounded; leaves scale-like, yellowish-green; seed cone berry-like, glaucous blue

CONIFEROUS TREES

Redwood and Sequoia

The towering redwoods and giant sequoias of California are not just botanical marvels; they are symbols of endurance, beauty, and the harmonious interplay between nature and time. These iconic species stand as living monuments, celebrating the state's natural heritage and capturing the imagination of millions who visit their ancient groves each year.

The redwood (*Sequoia sempervirens*) is the Earth's tallest tree species, reaching 381 feet. Giant sequoia (*Sequoiadendron giganteum*) claims the title of the most massive tree, by volume, in the world.

Ecologically, each species plays a critical role in their environments. Redwoods are vital in sequestering carbon, mitigating climate change by absorbing vast amounts of carbon dioxide. Giant sequoias are fire-adapted and depend on natural fire cycles to regenerate—however, with climate change, high-intensity fires are now a major threat to the species' survival. These trees act as keystone species in their ecosystems, fostering life and maintaining balance.

Moreover, redwoods and sequoias serve as powerful reminders of the importance of environmental stewardship. Their survival is a story of triumph and vulnerability. While these giants have endured for millennia, they face threats from climate change, logging, and human encroachment. California's commitment to protecting these forests ensures that future generations can continue to marvel at their grandeur and learn from their legacy.

Visiting an old-growth redwood or sequoia grove is a humbling experience. Walking among these ancient trees evokes a sense of awe and perspective, a reminder of humanity's place within the natural world. Their presence encourages us to think long-term, to consider the impact of our actions, and to strive for harmony with nature.

In celebrating the redwoods and giant sequoias, we honor more than their immense size or age; we recognize their role as enduring guardians of California's natural legacy. These trees inspire wonder, demand respect, and challenge us to protect the beauty and balance of our planet.

REDWOOD

Sequoia sempervirens

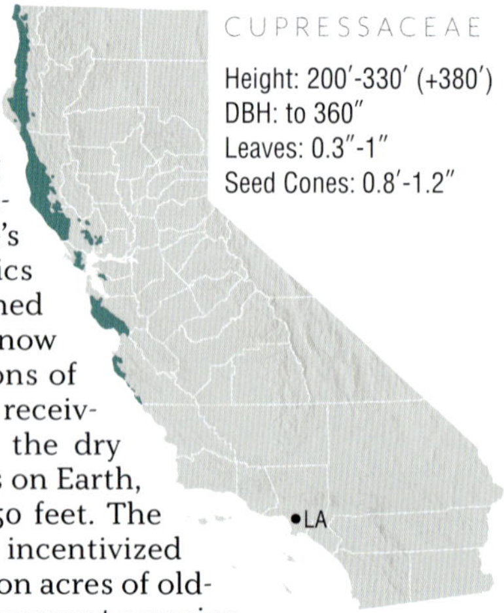

CUPRESSACEAE

Height: 200′-330′ (+380′)
DBH: to 360″
Leaves: 0.3″-1″
Seed Cones: 0.8′-1.2″

California's official state tree is among the world's most magnificent and revered organisms. Like many of the state's conifers, redwoods are surviving relics from a time when great forests flourished across the Northern Hemisphere. They now occur only in the mild climatic conditions of coastal central and northern California, receiving water from condensed fog during the dry summers. Redwoods are the tallest trees on Earth, with many individuals reaching over 350 feet. The economic value of the durable lumber incentivized ruthless logging of the original two million acres of old-growth stands, and now less than four percent remains. The oldest redwoods are estimated at 2,500 to 3,000 years old. The fire-resistant, cinnamon-colored bark is thick and spongy.

ID Tips: needles variable in length from bottom (longer) to top of tree (shorter); seed cones oblong-spherical; often dominant where they grow

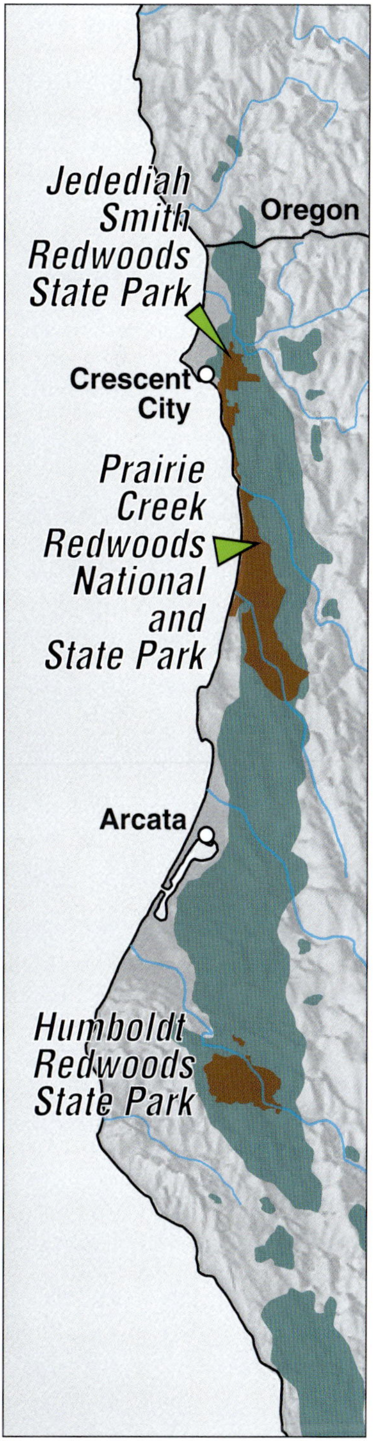
Jedediah Smith Redwoods State Park
Oregon
Crescent City
Prairie Creek Redwoods National and State Park
Arcata
Humboldt Redwoods State Park

GIANT SEQUOIA

Sequoiadendron giganteum

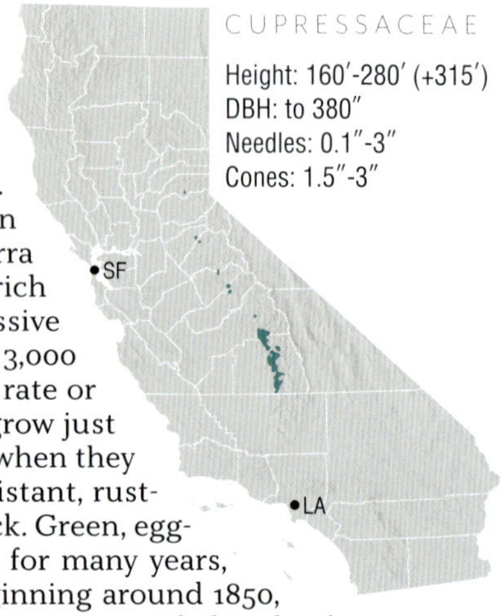

CUPRESSACEAE

Height: 160′-280′ (+315′)
DBH: to 380″
Needles: 0.1″-3″
Cones: 1.5″-3″

The massive and ancient giant sequoias are the sole members of their genus. They only grow natively in California in about 70 remaining groves in the Sierra Nevada, generally on moist, nutrient-rich soils. Giant sequoias are the most massive trees in the world, and some are over 3,000 years old. They don't decline in growth rate or reproductive output as they age. They grow just as fast when they are 2,500 years old as when they are 250 years old. Their spongy, fire-resistant, rust-colored bark can be up to three feet thick. Green, egg-shaped cones can remain on the trees for many years, growing as they age. For a century, beginning around 1850, trees were logged for grape stakes, fence posts, and shingles for towns and farms of the Central Valley. Most of the remaining groves are protected within national forests or national and state parks, though many trees have been lost recently to severe fires and drought.

ID Tips: leaves variable in length from bottom (longer) to top of tree (shorter); seed cones egg-shaped; dominant tree where they grow

A) Placer County Grove
B) Calaveras Big Trees
C) Mariposa Grove
D) Giant Forest
E) Freeman Creek (extirpated by fire)
A
South Lake Tahoe
Nevada
California
B
Yosemite National Park
Mariposa
C
Bishop
Sequoia-Kings Canyon National Park
D
Three Rivers
E
Giant Sequoia National Monument

Pine Family

The Pinaceae family has evolved mostly in the Northern Hemisphere since the Cretaceous (~140 Ma). Except for a few occurrences of one pine species, all family members are found north of the equator. The family in California includes true firs, spruces, pines, Douglas-firs, and hemlocks. It is characterized by needle-like leaves and large, woody cones. California's pine species, like the sugar pine and the ponderosa pine, grow to impressive heights and sizes. This family also includes the Great Basin bristlecone pine, known for its twisted, ancient trunks, and thriving in high-altitude environments in the eastern part of the state. True firs, such as white firs and red firs, are also members of this family, and define the snowy forests above 6,000 feet in elevation.

The pine family also has major economic importance, providing most of the softwood timber in North America. In the Northern Hemisphere, this family is the most ecologically important family of gymnosperm. Dispersed and planted by humans, species like Monterey pine have become naturalized in the Southern Hemisphere. Pinyon pine nuts are one of only a handful of commercially important foods from conifers (see p. 93).

Worldwide, this family has 11 genera with 230 species, representing over 35% of the world's conifers. In California, there are five genera and 32 species, about one-third of the tree species in the state! One must get to know this family to understand California's trees.

True Firs (pp. 68-75)

Spruces (pp. 76-80)

Pines (pp. 81-103)

Hemlocks (pp. 107-109)

Douglas-firs (pp. 104-106)

The pine family dominates the snow forests of California, above 5,000–7,000 feet depending on latitude, here with *Pseudotsuga menziesii* and *Abies concolor*

True Firs

Of the 50 species of fir worldwide, seven are native to California. Firs are difficult to distinguish, and many hybridize naturally. One of our firs is only found here (*Abies bracteata*), three are rare in California but more widespread elsewhere (*A. amabilis, A. lasiocarpa,* and *A. procera*), two are common in the high elevations of the Sierra and Cascades (*A. concolor* and *A. magnifica*), and one is found along California's northwestern coast (*A. grandis*). Firs have needle-like leaves borne singly, often curving upward along the stems. The upright seed cones are generally produced near the top of the tree and fall apart when mature. The needles' bases, which are shaped like suction cups, leave conspicuous circular scars when the leaves fall. This is a helpful feature to distinguish true firs from Douglas-firs.

Abies magnifica var. *shastensis* along the Pacific Crest Trail in the Marble Mountain Wilderness, Siskiyou County

PACIFIC SILVER FIR

Abies amabilis

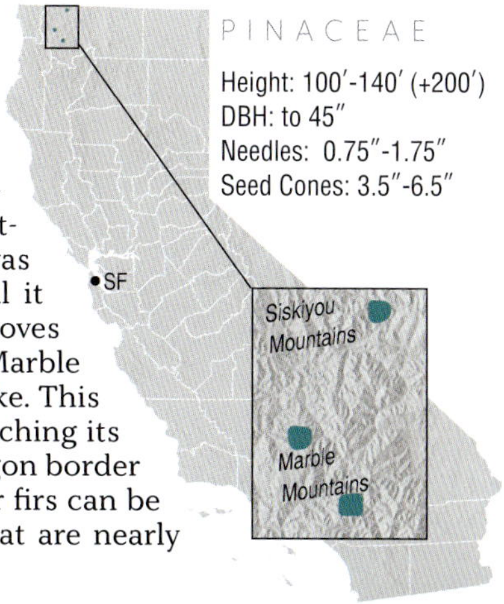

PINACEAE

Height: 100′-140′ (+200′)
DBH: to 45″
Needles: 0.75″-1.75″
Seed Cones: 3.5″-6.5″

Pacific silver fir, which gets its name from the silver undersides of the needles, is widespread in the cool, wet coastal areas of the Pacific Northwest. It was not known to occur in California until it was discovered in the 1930s in a few groves in western Siskiyou County, in the Marble Mountain Wilderness near Hancock Lake. This is one of California's rarest conifers, reaching its southern range extension near the Oregon border in the Klamath Mountains. Pacific silver firs can be massive, and individuals are known that are nearly 200 feet tall.

ID Tips: bark with rectangular blocks; needles dark green above, with bright white stomatal bloom on undersides; Klamath Mountains

SANTA LUCIA FIR

Abies bracteata

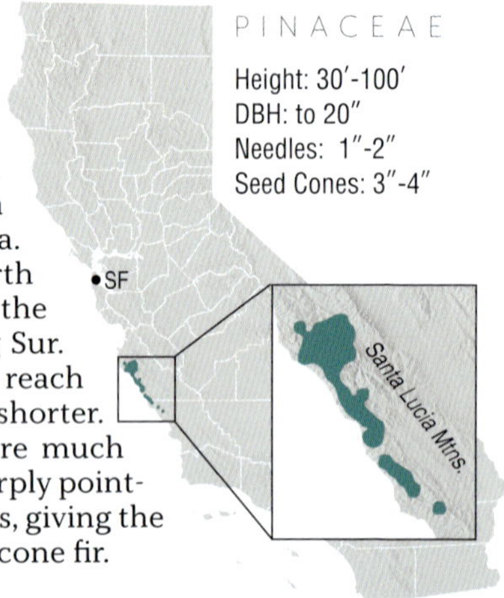

PINACEAE

Height: 30′-100′
DBH: to 20″
Needles: 1″-2″
Seed Cones: 3″-4″

Santa Lucia fir only grows within 15 miles of the coast in the Santa Lucia Mountains of central coastal California. It's the rarest and most unusual North American fir, growing in rocky soils on the steep slopes of fog-laden ravines in Big Sur. These narrow, spire-shaped trees can reach 100 feet in height, but most are much shorter. Compared to typical firs, their buds are much longer, their needles are much more sharply pointed, and their cones have long projections, giving the trees their other common name: bristlecone fir.

ID Tips: needle tips sharply pointed; seed cones with long extensions beyond the scales (bracts); Santa Lucia Mountains

WHITE FIR

Abies concolor

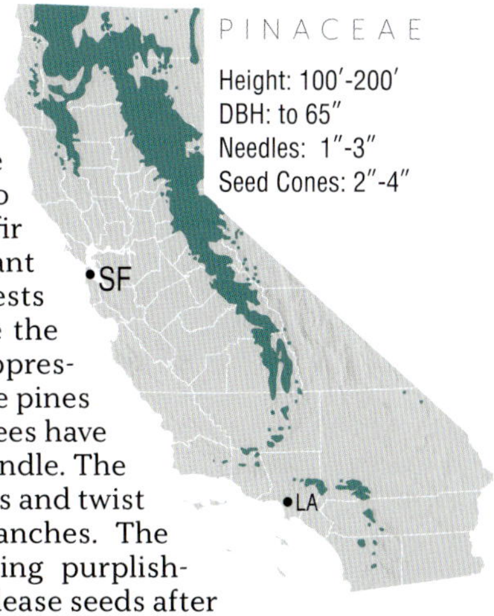

PINACEAE

Height: 100′-200′
DBH: to 65″
Needles: 1″-3″
Seed Cones: 2″-4″

White fir occurs widely throughout the southwestern U.S., from New Mexico to central Oregon. In California, white fir is the dominant and most shade-tolerant tree of the mid-elevation coniferous forests of the Sierra Nevada where they define the winter snowline. A century of fire suppression has allowed white fir to outcompete pines in many areas. The trunk bases of old trees have bark that looks like melting wax on a candle. The needles often lie flat on shaded branches and twist upward on higher, more exposed branches. The cones are green when young, becoming purplish-brown in age before disintegrating to release seeds after six to eight months.

ID Tips: stomatal bloom on underside of needles; gray-brown bark with dripping pattern at base; usually lower elevations than red fir

GRAND FIR

Abies grandis

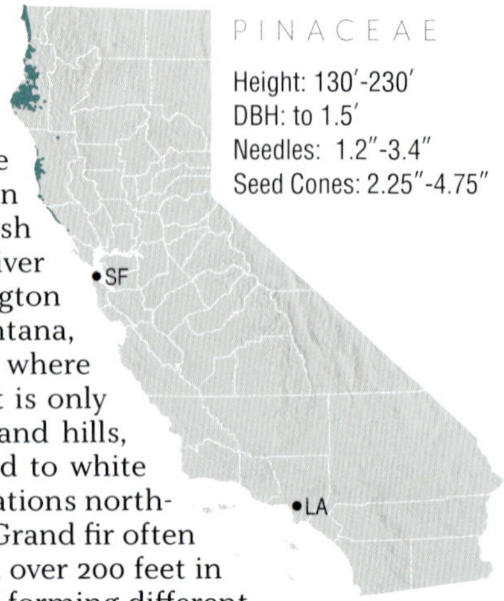

PINACEAE

Height: 130′-230′
DBH: to 1.5′
Needles: 1.2″-3.4″
Seed Cones: 2.25″-4.75″

Grand fir is widely distributed in the Pacific Northwest in western Oregon and Washington to southwestern British Columbia, and from the Columbia River basin of eastern Oregon and Washington to the northern Rockies of Idaho, Montana, and Canada. In California, however, where grand fir reaches its southern limits, it is only found growing on coastal headlands and hills, and in river valleys. It is closely related to white fir and hybridizes with it at higher elevations northward across Oregon and Washington. Grand fir often co-occurs with redwood and can reach over 200 feet in height. It has the interesting feature of forming different length needles along branchlets.

ID Tips: bark smooth with shallow furrows; needles often of two different lengths, with notches at tips; coastal forests below 2,000 feet

SUBALPINE FIR

Abies lasiocarpa

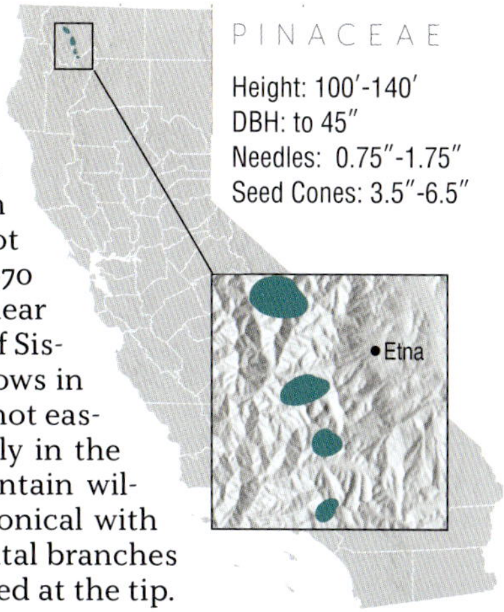

PINACEAE

Height: 100′-140′
DBH: to 45″
Needles: 0.75″-1.75″
Seed Cones: 3.5″-6.5″

Subalpine fir is the most widely distributed fir in North America. It occurs from northern Arizona to Alaska but was not known to occur in California until 1970 when it was found in a small stand near Russian Peak in the Salmon Mountains of Siskiyou County. Subalpine fir generally grows in small stands at higher elevations and is not easily observed in California, occurring only in the Russian, Trinity Alps, and Marble Mountain wilderness areas. The trees are narrowly conical with smooth trunks of gray bark and horizontal branches with flat needles that are blunt or notched at the tip.

ID Tips: distinctly tapering to pointed crown; bark smooth, even with age; usually in or near meadows

CONIFEROUS TREES

RED FIR

Abies magnifica

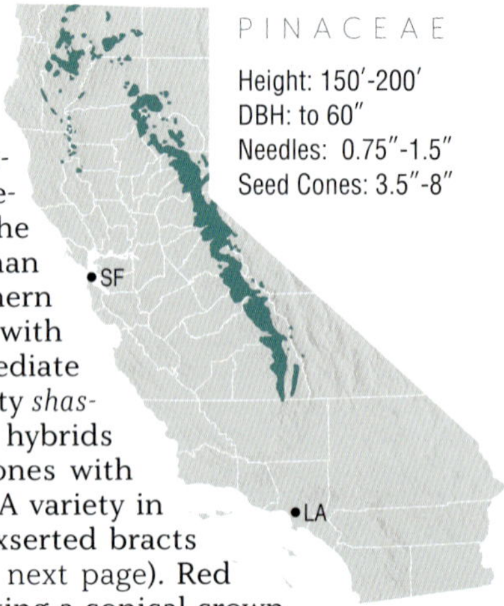

PINACEAE

Height: 150′-200′
DBH: to 60″
Needles: 0.75″-1.5″
Seed Cones: 3.5″-8″

In California, red fir dominates the high-elevation snowy forests of the Sierra Nevada, Klamath, and Cascade ranges. The range of red fir in these areas is wider than that of white fir (*A. concolor*). At the northern end of its distribution, red fir hybridizes with noble fir (*A. procera*), and these intermediate trees are sometimes treated as the variety *shastensis*, or Shasta red fir. These putative hybrids differ from typical red fir in having cones with partially exserted leaf-like projections. A variety in the southern Sierra Nevada also has exserted bracts (var. *critchfieldii*, see *A. procera* photo on next page). Red firs have reddish-brown trunks supporting a conical crown of bluish-green needles. Purplish-brown seed cones form high in the canopy and are the largest known of any fir species.

ID Tips: bark reddish, deeply furrowed; rust-colored hairy twigs, needles spreading in various directions, leaving stem with hockey-stick shape

NOBLE FIR

Abies procera

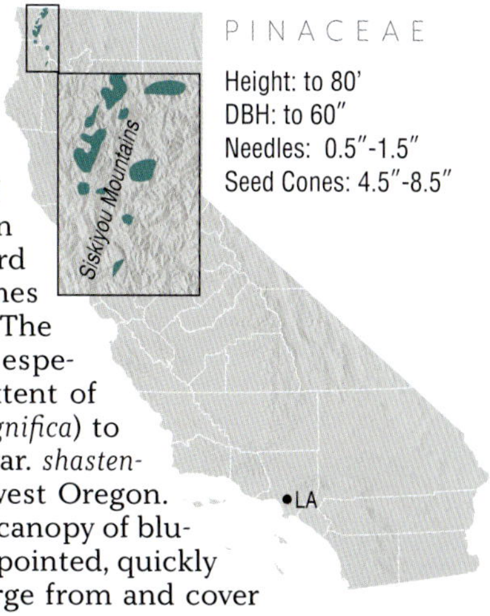

PINACEAE

Height: to 80'
DBH: to 60″
Needles: 0.5″-1.5″
Seed Cones: 4.5″-8.5″

In California noble fir is rare, growing grows only in the cool, moist forests of the Siskiyou Mountains, a coastal portion of the Klamath Mountains. Northward into Oregon and Washington it becomes the dominant tree of subalpine forests. The lines between species sometimes blur, especially in plants, and at the southern extent of noble fir intergrades with red fir (*A. magnifica*) to what is named Shasta fir (*A. magnifica* var. *shastensis*) in northwest California and southwest Oregon. Noble firs have a tightly packed conical canopy of bluish-green needles, and seed cones with pointed, quickly tapering leaf-like projections that emerge from and cover 90% of the cone surface.

ID Tips: crown tapering; bark smooth when young, furrows when mature; needles blue-green, spiral around stem; usually near meadows

Spruces

Three of the 34 species of the genus *Picea*, all of which occur in the Northern Hemisphere, grow in California. They are conical trees with four-sided, sharp-tipped needles that are borne singly in tight spirals along twigs. When the needles fall, they leave persistent woody bases on the twigs. The oval ovoid female cones dangle high in the crown, open and disperse seeds in autumn, and then fall to the ground intact.

Picea sitchensis decorate the coastal dune systems of Humboldt and Del Norte Co.

BREWER SPRUCE

Picea breweriana

PINACEAE

Height: 30′-120′ (+150′)
DBH: to 40″
Needles: 0.6″-1.4″
Seed Cones: 3″-6″

The rarest spruce in the U.S., Brewer spruce is scattered across the Klamath Mountains of northwestern California and southern Oregon, often occurring as isolated stands on high peaks. This long-lived species has a weeping canopy of dark green foliage and can reach heights of over 150 feet. The species has pendulous seed cones with bracts that are soft, but not as soft as the other local spruce species. Bark is light gray to brown with circular platelets that vary in color.

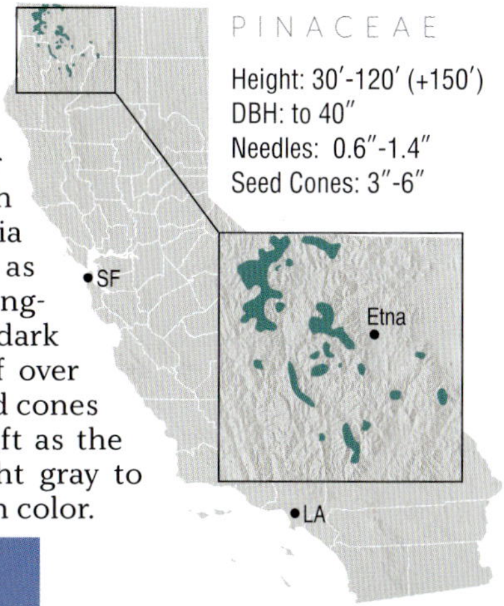

ID Tips: bark with smooth, gray to light brown platelets; branches drooping, hanging vertically from horizontal limbs

CONIFEROUS TREES

SITKA SPRUCE

Picea sitchensis

PINACEAE

Height: to 300′ (+329′)
DBH: to 190″
Needles: 0.8″-1.2″
Seed Cones: 2.5″-4″

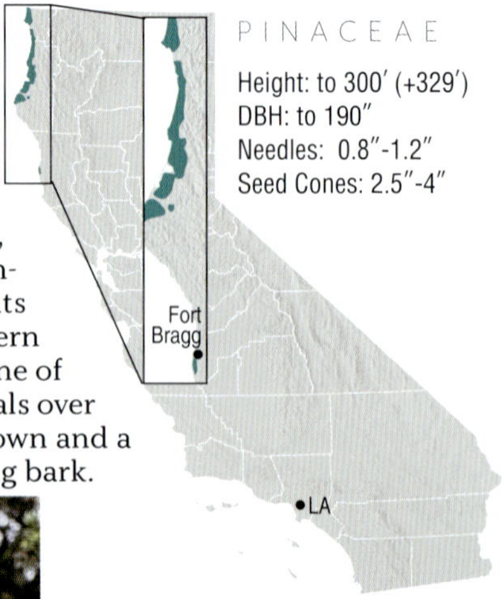

Sitka spruce reaches the southern end of its 2,000-mile range on the coast in Mendocino County. It is a coastal tree, rarely occurring more than 30 miles inland and grows in wet or boggy habitats from northern California to southern Alaska. The largest of all spruces, it is one of the few trees in the world with individuals over 300 feet tall. It has a narrow, conical crown and a buttressed trunk covered in thin, flaking bark.

ID Tips: bark grayish-brown with round scales; needles single with blue-gray tint and sharp points; seed cones thin, light brown, with papery scales

The World's Tallest Trees

These are remarkable examples of nature's grandeur. Growing in optimal conditions around the Pacific Rim, each species depends on specific ecological conditions to reach these heights including moisture, soil quality, and climate stability (lack of hurricane-force winds!). **Three of the seven tallest tree species on Earth grow in California!**

Seven Tallest Tree Species[1]	Height	Type	Location
Redwood (*Sequoia sempervirens*)	381′	Conifer	**California**
Tibetan cypress (*Cupressus austrotibetica*)	335′	Conifer	Tibet
Mountain ash[2] (*Eucalyptus regnans*)	330′	Flowering Tree	Tasmania
Sitka spruce (*Picea sitchensis*)	329′	Conifer	**California**
Douglas-fir[3] (*Pseudotsuga menziesii*)	326′	Conifer	Oregon*
Yellow meranti (*Shorea faguetiana*)	326′	Flowering Tree	Borneo
Giant sequoia (*Sequoiadendron giganteum*)	316′	Conifer	**California**

1. Measured since 2001
2. Height changes; the *E. regnans* lost part of its crown and is 316′ in 2025
3. There is a 324′ Douglas-fir in California

ENGELMANN SPRUCE

Picea engelmannii

PINACEAE

Height: 80′-120′
DBH: to 50″
Needles: 0.6″-1″
Seed Cones: 1″-3″

Engelmann spruce grows from the Chiricahua Mountains of Arizona to British Columbia, but it is rare in California, occurring in only a few stands in the Klamath Mountains, centered in the Russian Wilderness, and in the northern California Cascades on Clark's Creek near Burney. They are tall, conical conifers with dark green foliage, thin, flaking bark, and light brown cones that hang high in the canopy.

ID Tips: trunk bases buttressed; needles sharp, stiff, square-shaped, forming a spiral around stem; seed cones papery, pendent

Pines

The 120 species of pine dominate vast areas of forested land in the Northern Hemisphere. In California, there are more pine species than any other type of tree, 18 total, and three (*Pinus balfouriana, P. sabiniana,* and *P. torreyana*) that grow only in the state. The slender evergreen needles grow in bundles of one to five, depending on the species. Small pollen-producing cones and larger, woody, many-scaled seed-producing cones occur on the same tree. Seed cones generally mature in two years, with two winged seeds on each cone scale.

Five-needle Pines

(pp. 82-90)

whitebark, foxtail, Great Basin bristlecone, western white, sugar, limber, and Torrey

One-, two-, and four-needle pines

(pp. 91-94)

Colorado pinyon, singleleaf pinyon, and Parry pinyon

Two- or three-needle closed-cone pines

(pp. 95-97)

knobcone, bishop, and Monterey

Two-needle pines

(pp. 98-99)

lodgepole and beach

Three-needle pines

(pp. 100-103)

ponderosa, Jeffrey, Coulter, and ghost

WHITEBARK PINE

Pinus albicaulis

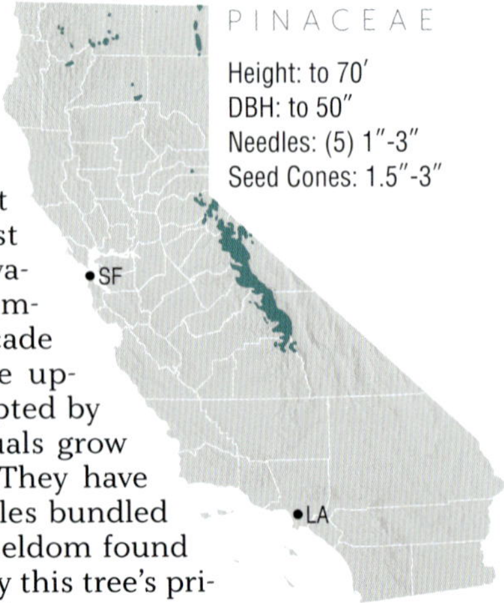

PINACEAE

Height: to 70′
DBH: to 50″
Needles: (5) 1″-3″
Seed Cones: 1.5″-3″

This is often the highest-elevation tree on mountainsides. In California, it occurs primarily along the jagged crest of the Sierra Nevada. With enough elevation, the species is sporadic in the Klamath Mountains and the southern Cascade Range. It survives near treeline, at the upper edge of where trees can grow. Sculpted by winds and extreme snowpack, individuals grow wind-shorn and close to the ground. They have chalky white bark and dark green needles bundled in fives. Intact mature seed cones are seldom found because they are torn apart and eaten by this tree's primary seed disperser, the Clark's nutcracker.

ID Tips: 5 needles per bundle; bark thin, scaly, grayish; branches often sculpted by harsh conditions at high elevations

CONIFEROUS TREES

FOXTAIL PINE

Pinus balfouriana

PINACEAE

Height: to 90′
DBH: to 100″
Needles: (5) 1″-3″
Seed Cones: 2.5″-4.5″

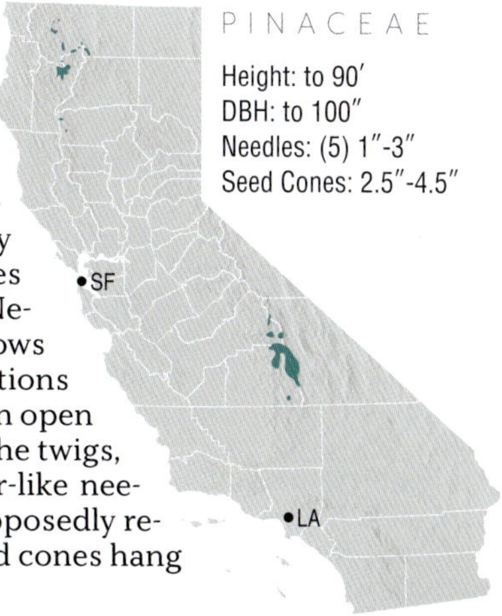

Foxtail pine, a five-needle pine found only in California, has two local populations (each a different subspecies) widely separated by over 300 miles; subspecies *austrina* occurs in the southern Sierra Nevada, whereas subspecies *balfouriana* grows in the Klamath Mountains. Both populations have long-lived, scattered trees growing in open subalpine stands on bare rocky slopes. The twigs, covered evenly with tufts of coarse, hair-like needles, are stout and slightly drooping, supposedly resembling a fox's tail. The egg-shaped seed cones hang from branches on short stalks.

ID Tips: 5 needles per bundle; bark gray in Klamath Mountains, red in Sierra Nevada; highest mountain summits

CONIFEROUS TREES

GREAT BASIN BRISTLECONE PINE

Pinus longaeva

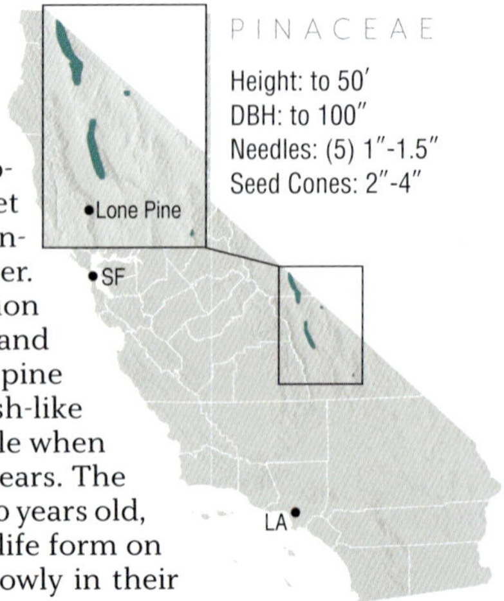

PINACEAE

Height: to 50′
DBH: to 100″
Needles: (5) 1″-1.5″
Seed Cones: 2″-4″

Bristlecone pines grow in scattered subalpine groves, generally above 10,000 feet in the Panamint, Inyo, and White mountains near California's eastern border. They range eastward atop high-elevation peaks of the Great Basin in Nevada and Utah. Bristlecone pine is a five-needle pine with needles clustered on bottlebrush-like branchlets. Their bristly cones are purple when young and mature to brown after two years. The oldest bristlecone pine approaches 5,000 years old, making it the oldest known non-clonal life form on Earth. These ancient trees grow very slowly in their cold, dry, and windswept environment.

ID Tips: 5 needles per bundle, packed tightly around stem; bark thin, smooth, furrowing and red-brown in age; high desert mountains

World's Oldest Living Thing?

The Great Basin bristlecone pine (*Pinus longaeva*), particularly the individual known as "Methuselah" in California's White Mountains, is one of the world's oldest known living trees, with an age of over 4,800 years. However, whether it is the "world's oldest living thing" depends on how "oldest" and "living" are defined.

Contenders for Oldest Living Organisms

Clonal organisms

- **Quaking aspen (*Populus tremuloides*)** A clonal colony in Utah is estimated to be about 80,000 years old. The individual trees are genetically identical stems connected by a single root system.
- **Neptune grass (*Posidonia oceanica*)** A seagrass meadow in the Mediterranean Sea is potentially tens of thousands of years old.

Non-clonal organisms

- **Great Basin bristlecone pine** "Methuselah" and other unnamed specimens of the same species hold the record for single, non-clonal living organisms approaching 5,000 years.

Microbial life

- **Certain bacteria and archaea** in extreme environments, like deep-sea vents or Antarctic permafrost, may have survived for millions of years, though these are not "individuals" in the same sense as trees are.

So while the bristlecone pine is most likely the oldest individual, non-clonal tree known, it's not necessarily the "oldest living thing" when considering clonal colonies or microbial life.

WESTERN WHITE PINE

Pinus monticola

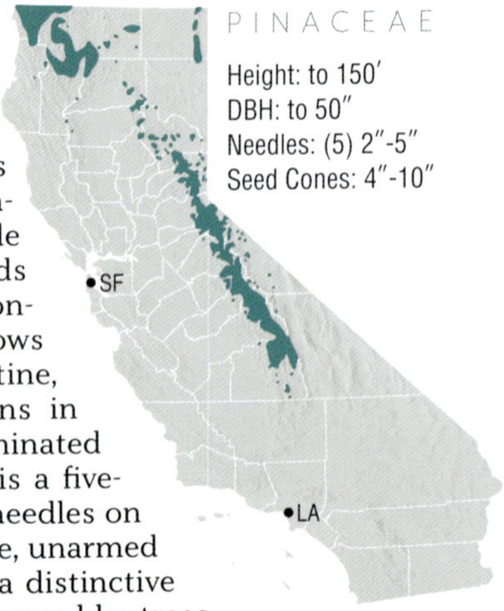

PINACEAE

Height: to 150′
DBH: to 50″
Needles: (5) 2″-5″
Seed Cones: 4″-10″

In California, western white pine grows from the southern Sierra Nevada northward through the Klamath and Cascade ranges. Outside the state, its range extends into Oregon, Washington, Idaho, Montana, and southwestern Canada. It grows on various rock types, including serpentine, and generally occurs at high elevations in scattered populations in habitats dominated by other conifers. Western white pine is a five-needle pine with slender, bluish-green needles on branches that swoop upward. Finger-like, unarmed cones hang from the branch tips, and a distinctive reddish-brown, checkered bark develops on older trees.

ID Tips: 5 needles per bundle; bark of large trees hexagonally checked, reddish in eastern California, grayish in eastern Klamath Mountains

SUGAR PINE

Pinus lambertiana

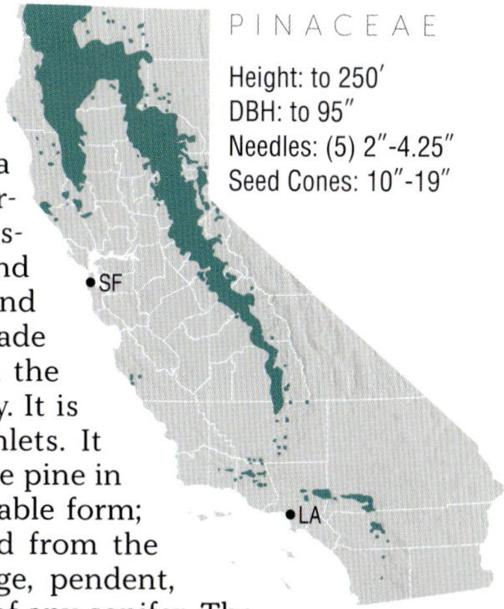

PINACEAE

Height: to 250′
DBH: to 95″
Needles: (5) 2″-4.25″
Seed Cones: 10″-19″

Sugar pine ranges from northern Baja California, Mexico to Oregon. In California, it grows in the Peninsular and Transverse ranges of southern California and northward through the Sierra Nevada and the North Coast, Klamath, and Cascade ranges, with an outlying population in the South Coast Range in Monterey County. It is a five-needle pine with slender branchlets. It grows to be the tallest and most massive pine in the world. Sugar pines have a recognizable form; the branches sweep gracefully outward from the upper trunk, bearing at their tips large, pendent, unarmed cones, which are the longest of any conifer. The common name comes from a sweet resin that hardens around fire scars.

ID Tips: 5 needles per bundle; bark furrowed into linear blocks, reddish-brown; common in mid-elevation (4,000–8,000 ft) conifer forests

LIMBER PINE

Pinus flexilis

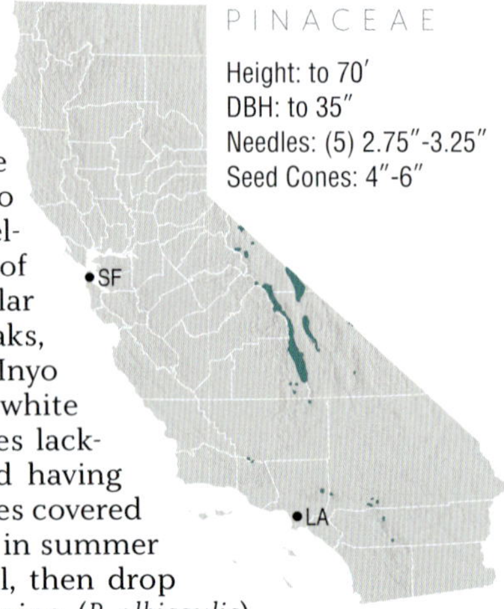

PINACEAE

Height: to 70′
DBH: to 35″
Needles: (5) 2.75″-3.25″
Seed Cones: 4″-6″

Limber pine occurs throughout the Rocky Mountains from northern Mexico to Alberta, Canada, generally at high elevations. It reaches the western edge of its range in California in the Peninsular and Transverse ranges, on desert peaks, and in the Sierra Nevada, White, and Inyo mountains. It is one of California's six white pines, each with five-needle leaf bundles lacking a sheath of scales at the base and having cones without a prickle. Flexible branches covered in grayish-brown bark bear green cones in summer that mature to yellow-brown in the fall, then drop seeds after two years. Like whitebark pine (*P. albicaulis*), limber pine also co-evolved with Clark's nutcrackers (*Nucifraga columbiana*) which tear open cones and eats, caches, and dispersers the tree's seeds.

ID Tips: 5 needles per bundle; bark furrowed into linear blocks, grayish-brown; high mountains

Threats to our 5-needle Pines

All is not well with the five-needle pines of western North America. Climate change and human-caused disturbances are reshaping forests and mountains. Water in the mountains is drying up earlier each season. Snowpack levels are decreasing. Drought conditions and warmer annual average temperatures are driving species typically found at lower elevations upward. Among the five-needle pines, the foxtail and whitebark pines are particularly impacted as different species migrate higher, bringing with them pathogens that the high-elevation trees don't have many defenses against.

Join the community science project to track 5-needle pines

DWARF MISTLETOES

Even though *Arceuthobium campylopodum* is native and has co-evolved with 5-needle pines, with drought and climate change dwarf mistletoe is having more drastic impacts on tree health.

BARK BEETLES

Bark beetles generally kill larger trees with variable outbreak size. They are becoming more common at higher elevations as the climate warms and beetles are able to reproduce more often at higher elevations.

WHITE PINE BLISTER RUST

This non-native rust fungus infects and kills various 5-needle pines. Symptoms include branch dieback and slow tree decline. It is becoming more common in California.

TORREY PINE

Pinus torreyana

PINACEAE

Height: to 55′
DBH: to 25″
Needles: (5) 10″-12″
Seed Cones: 3″-6″

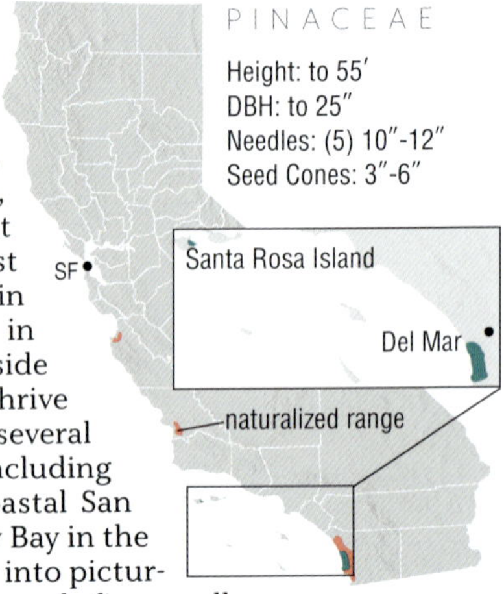

Torrey pine, occurring naturally only in two groves in southwestern California, is the rarest pine in the U.S. Only about 7,000 wild individuals grow along the coast of northern San Diego, almost entirely in Torrey Pines State Natural Reserve and in a small grove on Santa Rosa Island. Outside their narrow native range, these trees thrive in cultivation and have naturalized in several areas along California's central coast, including northwestern Santa Barbara County, coastal San Luis Obispo County, and along Monterey Bay in the town of Marina. Torrey pines grow slowly into picturesque, broad-canopied trees. They are the only five-needle pines in California with long needles. Like Coulter pine (*P. coulteri*) and foothill pine (*P. sabiniana*), Torrey pines have long, gray-green needles and large seed cones that can remain on the tree for years.

ID Tips: 5 very long-needles per bundle; large seed cones with stout prickles on scales; bark with dark-gray plates that flake to reveal a red-brown

CONIFEROUS TREES

COLORADO PINYON PINE

Pinus edulis

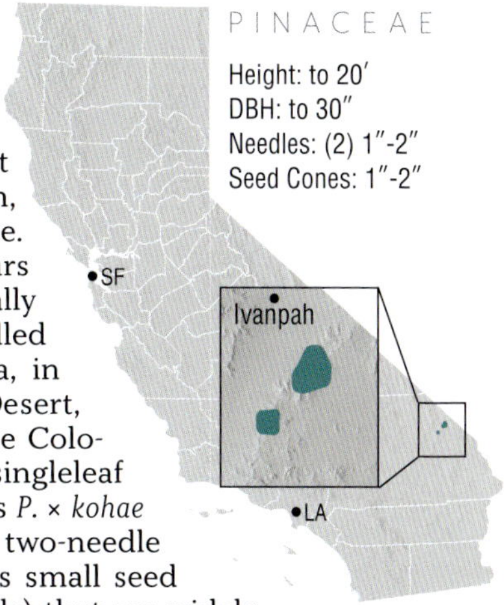

PINACEAE

Height: to 20′
DBH: to 30″
Needles: (2) 1″-2″
Seed Cones: 1″-2″

Colorado pinyon is rare in California but common in Arizona, New Mexico, Utah, and Colorado—where it is the state tree. Throughout its extensive range, it occurs with several species of juniper, especially Utah juniper, in a dry, rocky habitat called pinyon-juniper woodland. In California, in the mountains of the eastern Mojave Desert, at the far western edge of its range, the Colorado pinyon forms hybrids with the singleleaf pinyon (*P. monophylla*). Some treat this as *P.* × *kohae* but we choose to treat it as *P. edulis*. It is a two-needle pine with short, dark green needles. Its small seed cones contain delicious pine nuts (seeds) that are widely collected and eaten by wildlife and humans alike.

ID Tips: 2 needles per bundle; bark scaly, rough, reddish-brown; seed cones egg-shaped; New York Mountains

SINGLELEAF PINYON PINE

Pinus monophylla

PINACEAE

Height: to 60′
DBH: to 30″
Needles: (1) 1.5″-2.25″
Seed Cones: 1.75″-3″

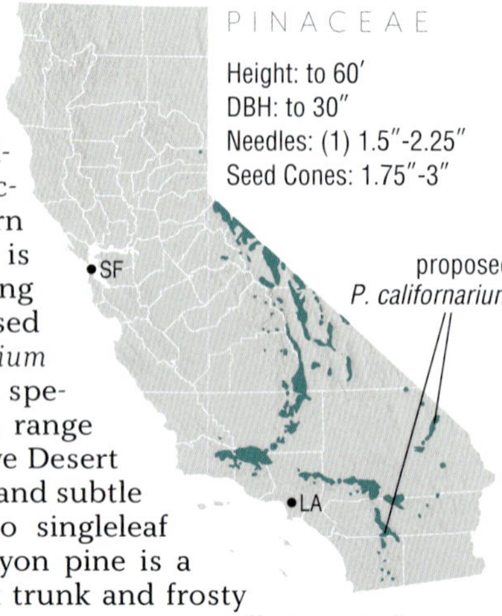

The pinyon pines (sometimes spelled piñon) are related pines that primarily occur in the southwestern U.S. and northern Mexico. The taxonomy of these species is in flux, with new genetic data shedding light on the dynamics of evolution based on isolation and habitat. *Pinus californarium* (not treated here) is a newly-proposed species described within the distribution range of *Pinus monophylla* in California's Mojave Desert with an exceptional drought tolerance (and subtle visible differences) when compared to singleleaf pinyon. As treated here, singleleaf pinyon pine is a small, broad-crowned tree with a black trunk and frosty gray to blue-green foliage. This species is unique among all pines in having single needles, whereas all others have needles bundled in groups of two to five. Each unarmed cone produces large, edible seeds (pine nuts).

ID Tips: 1 needle per bundle, only pine with single needles; needles thick, sharp-tipped, bluish-green; seed cones egg-shaped

Edible Conifers?

Many conifers produce edible components that have sustained humans for millennia. The seeds of pines are a notable example, with species like the pinyon pines of North America and stone pines of Europe prized for their nutrient-dense seeds (pine nuts). Beyond seeds, the inner bark (cambium) of some conifers, such as the Norway spruce and lodgepole pine, has historically been consumed in times of scarcity and used as flour. Young shoots, needles, and pollen from various species, like fir and spruce, are also edible, often used for teas, syrups, or as nutritional supplements rich in vitamin C. However, not all conifers are safe to eat; some, such as yew, are toxic, requiring careful identification before consumption.

Pinyon pine nut harvesting in California has a rich history that spans thousands of years. First Peoples, such as the Piute, Shoshone, and other Indigenous groups, gather pine nuts from pinyon pine trees (primarily the singleleaf pinyon). These nuts were a cornerstone of their diet and culture, providing a crucial source of calories and nutrients.

Pre-1850, First Peoples used specialized tools like long poles to knock cones from trees or climbed trees to collect them. The sustainable harvest was communal and tied to seasonal migrations and ceremonies. To ensure the nut's preservation for storage and transport, they were dried and left in the shell or ground into flour. These nuts are high in calories, healthy unsaturated fats, protein, and vitamins.

Today, pine nut harvesting persists both as a cultural practice among First Peoples and as a small-scale commercial or recreational activity. In some areas, commercial harvesting has raised concerns about sustainability and competition with wildlife like squirrels and birds. Harvesting pine nuts often requires a permit depending on the location and intended use.

Pinus monophylla seeds (pine nuts) both encased and not

PARRY PINYON PINE

Pinus quadrifolia

PINACEAE

Height: to 50′
DBH: to 15″
Needles: (4) 1″-2″
Seed Cones: 1.5″-2.25″

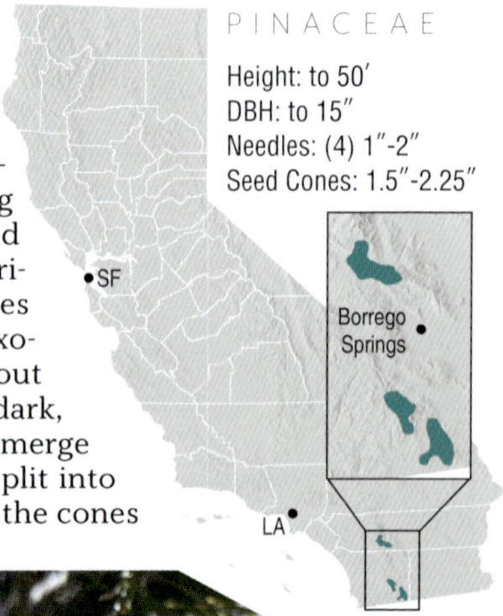

Parry pinyon pine grows in open woodlands and chaparral on desert-facing slopes of the Peninsular Ranges and southward into Baja California. This variable species, generally with four needles per cluster, has been the subject of taxonomic controversy. Parry pinyons are stout trees, with bluish-green foliage, and dark, scaly bark. The needles of new growth emerge seemingly as a single unit and quickly split into four. As with other pinyon pine species, the cones have edible seeds.

ID Tips: 4 needles per bundle (generally); bark thick, scaly, reddish-brown, becoming deeply furrowed with age; seed cones small, globose

KNOBCONE PINE

Pinus attenuata

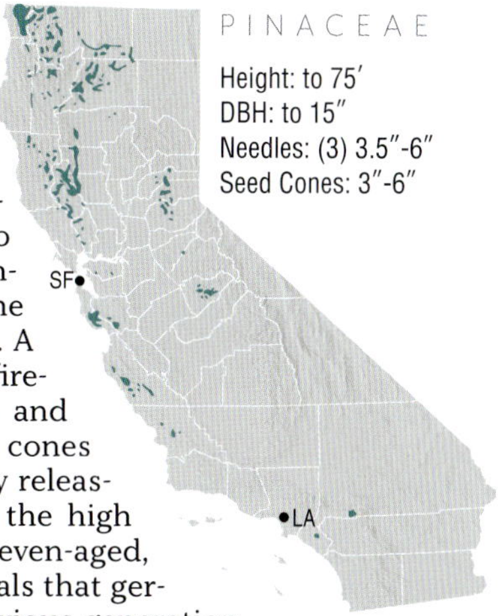

PINACEAE

Height: to 75′
DBH: to 15″
Needles: (3) 3.5″-6″
Seed Cones: 3″-6″

Knobcone pine occurs discontinuously on arid, rocky slopes and ridges in coastal ranges from San Bernardino County to southwestern Oregon and from the central Sierra Nevada northward through the Klamath Mountains and Cascade Range. A closed-cone pine, it is California's most fire-dependent conifer. Along with bishop and Monterey pines, it has thick, woody seed cones whose scales stay sealed with pitch, only releasing their seeds after being exposed to the high temperatures of a fire. Most stands are even-aged, often forming dense thickets of individuals that germinated after the fire that killed the previous generation. Knobcone pine is a three-needle pine with wispy crowns of pale green foliage and slender trunks and branches covered with long-persistent, asymmetric seed cones.

ID Tips: 3 needles per bundle; seed cones asymmetric, remaining on trunk and branches; trees relatively small, scruffy, in even-aged stands

CONIFEROUS TREES

BISHOP PINE

Pinus muricata

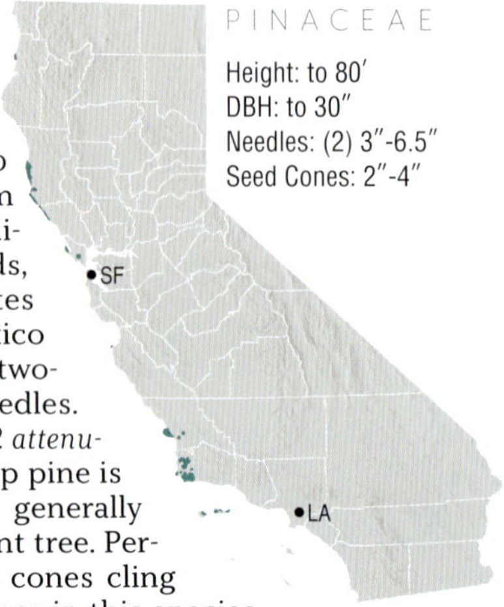

PINACEAE

Height: to 80′
DBH: to 30″
Needles: (2) 3″-6.5″
Seed Cones: 2″-4″

Bishop pine always grows within 10 miles of the Pacific Ocean, ranging from Cedros Island off the coast of Baja California to the California Channel Islands, and on the mainland in scattered sites from northwestern Baja California, Mexico to Humboldt County. Bishop pine is a two-needle pine with stiff, often twisted needles. Like its close relatives, knobcone pine (*P. attenuata*) and Monterey pine (*P. radiata*), bishop pine is a closed-cone species with cones that generally open only after a fire has killed the parent tree. Persistent, unopened, usually asymmetric cones cling in tight whorls around the branches. Trees in this species are highly variable, becoming dwarfed on poorly drained, nutrient-deprived soils or growing as tall trees in dense forests in protected habitats. On headlands, trees are broad and wind-shorn with grayish-brown furrowed bark.

ID Tips: 2 needles per bundle, needles stiff, twisted; seed cones like Monterey pine but smaller and with sharp prickles; grows close to the ocean

MONTEREY PINE

Pinus radiata

PINACEAE

Height: to 90′
DBH: to 30″
Needles: (3) 3″-6″
Seed Cones: 4″-6.5″

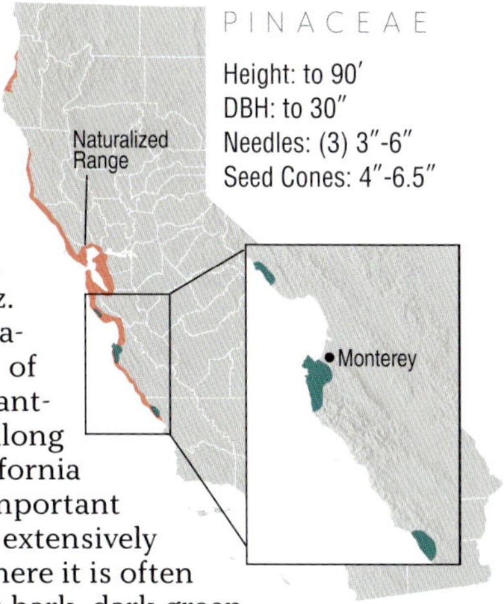

Monterey pine is rare in the wild, occurring only in three isolated mainland stands along the California coast: in Cambria, on the Monterey Peninsula, and along the coast north of Santa Cruz. Small populations also exist on Guadalupe and Cedros islands off the coast of Baja California, Mexico. It has been planted as an ornamental in many areas along the California coast. Away from its California home, it is one of the world's most important plantation timber trees, having been extensively planted in the Southern Hemisphere, where it is often weedy. Monterey pine has grayish-black bark, dark green needles generally in clusters of three, and knobby, light brown, asymmetric cones that persist on the trunk and branches. Although it is a closed-cone species like its relatives knobcone pine (*P. attenuata*) and bishop pine (*P. muricata*), it takes less heat to open the cones of Monterey pine. Sometimes, on a hot day, a loud crack signals that a cone has popped open.

ID Tips: 3 needles per bundle (sometimes 2 in cultivation); bark red- to blackish-brown, deeply furrowed; seed cones egg-shaped on branches

LODGEPOLE PINES

Pinus contorta subspecies

Lodgepole pines are two-needle pines with a wide distribution in western North America, occupy a greater elevational range and broader range of habitats than any other pine species. Three of the four taxa (3 subspecies, one variety) grow in California. Sierra lodgepole pine (subsp. *murrayana*) occurs at high elevations, often in wetter areas in the eastern Transverse, Sierra Nevada, Inyo, White, Klamath, Cascade, and Warner ranges. It is generally an erect tree with thin, flaky bark with golf-ball-sized seed cones that open when mature. Shore pine (subsp. *contorta*) often has crooked trunks with thick, furrowed bark, and grows on coastal terraces from northern California to Alaska. The third California two-needle pine, the Bolander's pine (variety *bolanderi*), is a dwarf tree constrained to stunted forests on highly acidic soils of old marine terraces around Fort Bragg in Mendocino County. It has closed cones that release seeds only after fire. Rocky Mountain lodgepole pine (subsp. *latifolia*) doesn't occur in California.

SIERRA LODGEPOLE PINE

Pinus contorta subsp. *murrayana*

ID Tips: 2 needles per bundle; bark scaly, not evidently furrowed, orange- to purple-brown; high-elevation mountains

CONIFEROUS TREES

PINACEAE

Height: to 100′ (*murrayana*) & to 40′ (*contorta*)
DBH: to 35″ (*murrayana*) & to 20″ (*contorta*)
Needles: (2) 1.2″-2.5″
Cones: 1.5″-2.5″

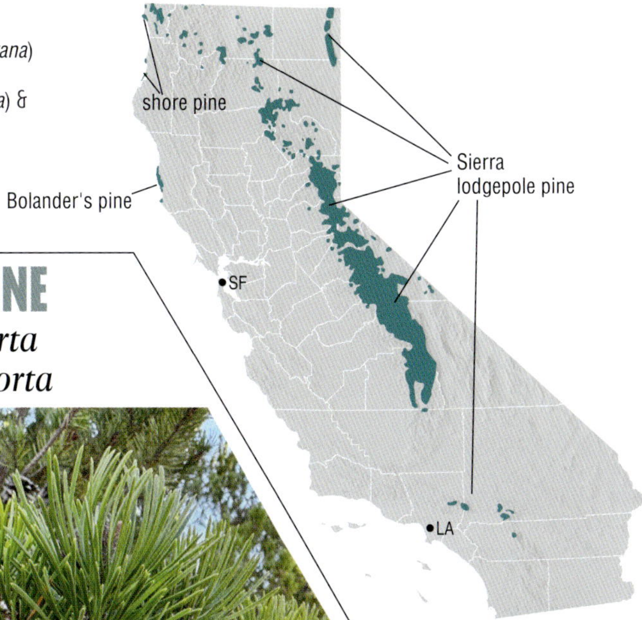

SHORE PINE

Pinus contorta subsp. *contorta*

ID Tips: 2 needles per bundle; bark furrowed into blocks; within a mile of Pacific Ocean, often in dunes

PONDEROSA PINE

Pinus ponderosa

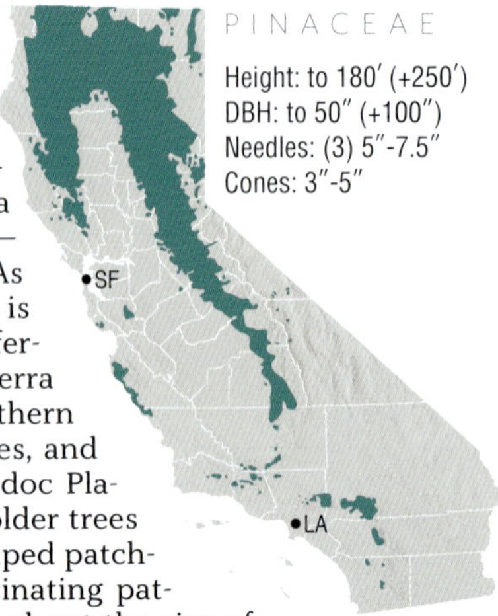

PINACEAE

Height: to 180′ (+250′)
DBH: to 50″ (+100″)
Needles: (3) 5″-7.5″
Cones: 3″-5″

Ponderosa pine, also called western yellow pine, is a three-needle pine with a wide distribution in the American west—growing in a diverse range of habitats. As California's most widespread conifer, it is abundant in low- to mid-elevation coniferous forests on the western side of the Sierra Nevada, in the mountain ranges of southern California, throughout the coastal ranges, and into northeastern California on the Modoc Plateau. The fissured and cracked bark of older trees forms orangish-brown jigsaw puzzle-shaped patches, each tree having a distinct and fascinating pattern and smells of vanilla. Prickly cones about the size of a baseball grow on branch tips bearing rough, aromatic needles. Trees have symmetric, pyramidal crowns and can reach heights of over 200 feet.

ID Tips: 3 needles per bundle; seed cone tips point inward, therefore are soft to the touch; often on poorer soils than ponderosa pine

JEFFREY PINE
Pinus jeffreyi

PINACEAE

Height: to 130′
DBH: to 35″
Needles: (3) 5″-9″
Seed Cones: 5.5″-9.5″

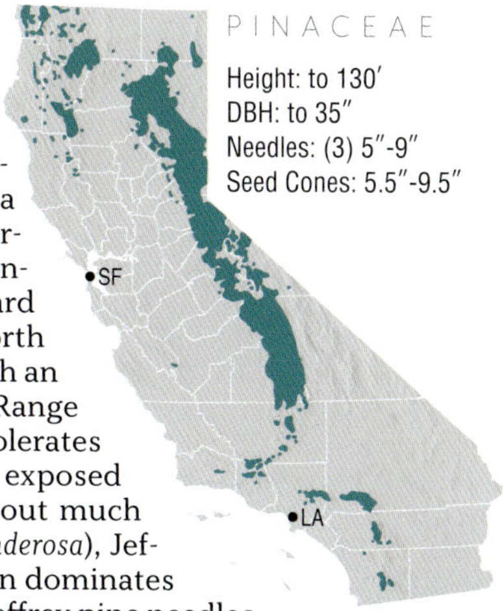

Jeffrey pine occurs primarily in California, from the mountains of northern Baja California, Mexico to southwestern Oregon. In California, it occurs from the Peninsular and Transverse ranges northward through the Sierra Nevada to the North Coast, Klamath, and Cascade ranges, with an outlying population in the South Coast Range in San Benito County. Jeffrey pine tolerates extreme conditions and grows in dry, exposed habitats, including serpentine, throughout much of the state. Like ponderosa pine (*P. ponderosa*), Jeffrey pine is a three-needle pine that often dominates California's montane forests. However, Jeffrey pine needles have a duller green color than those of ponderosa pine, and its cones are larger with an incurved prickle on the tip of each cone scale. Its bark releases a sweet aroma of butterscotch.

ID Tips: 3 needles per bundle; seed cone tips point inward, therefore are soft to the touch; often on poorer soils than ponderosa pine

COULTER PINE

Pinus coulteri

PINACEAE

Height: to 75′
DBH: to 36″
Needles: (3) 6″-12″
Seed Cones: 8″-16″

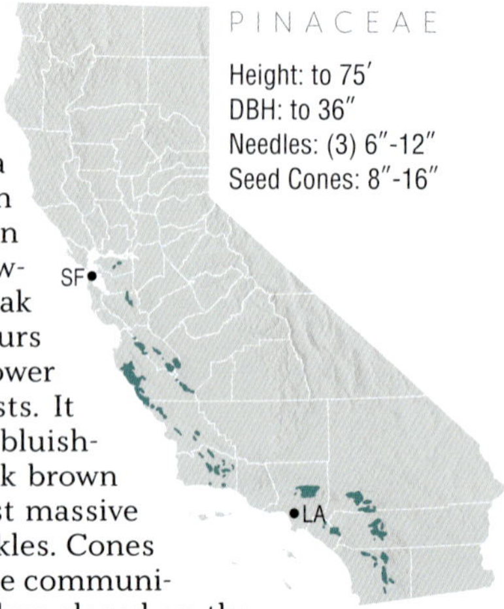

Coulter pine ranges from northern Baja California, Mexico to the eastern San Francisco Bay Area. It does not occur in the interior mountains of the state. At lower elevations, it grows in chaparral and oak woodlands; at higher elevations, it occurs from open, dry woodlands into the lower elevations of montane coniferous forests. It is a three-needle pine with long, stiff, bluish-gray needles and roughly furrowed, dark brown bark. The heavily armed cones, the most massive of any pine, have curved, talon-like prickles. Cones in populations associated with fire-prone communities, such as chaparral, persist more or less closed on the branches, whereas cones on trees in less fire-prone areas open and fall when mature.

ID Tips: 3 needles per bundle, glaucous gray-green; seed cones exceptionally large, heavy, prickles, averaging 12 inches long

GHOST PINE

Pinus sabiniana

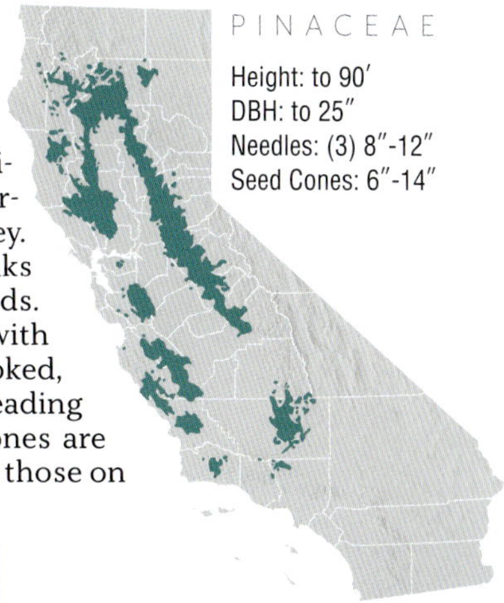

PINACEAE

Height: to 90′
DBH: to 25″
Needles: (3) 8″-12″
Seed Cones: 6″-14″

Ghost pine grows natively only in California, roughly in an elongated ring surrounding California's Great Central Valley. This species often grows with blue oaks (*Quercus douglasii*) in foothill woodlands. It is a long-needled, three-needle pine with sparse, bluish-gray foliage. The crooked, black-barked trunk bears irregular, spreading to upright branches. The large seed cones are armed with stout, recurved prickles like those on Coulter pine (*P. coulteri*).

ID Tips: 3 needles per bundle, silvery-gray (ghostly appearance); crown open, irregular; bark furrowed, dark; seed cones with sharp scales

Douglas-firs

Of the six species of *Pseudotsuga*, four are from Asia, and two grow in California. The distinctive cones are pendent, with prominent, three-pointed bracts emerging from scales, giving the appearance of the tail end of a mouse dashing into the cone.

A massive *Pseudotuga menziesii* grows along Eagle Creek in the mixed-evergreen forest of the Trinity Alps Wilderness, Trinity County

DOUGLAS-FIR

Pseudotsuga menziesii

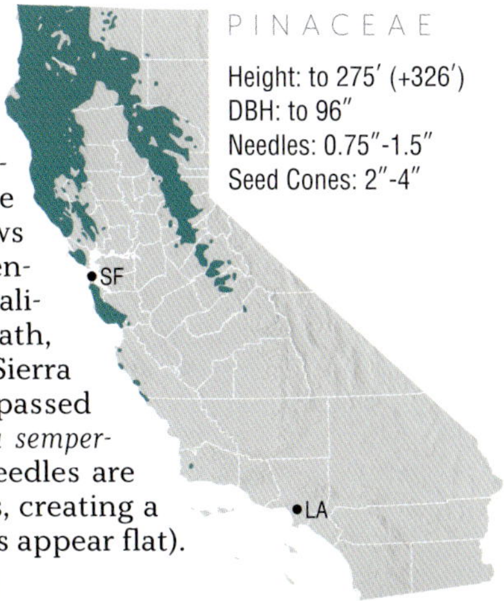

PINACEAE

Height: to 275′ (+326′)
DBH: to 96″
Needles: 0.75″-1.5″
Seed Cones: 2″-4″

Douglas-fir is North America's most important timber producer and one of the world's most impressive trees. It grows throughout the western U.S., from central California to British Columbia. In California, it covers the North Coast, Klamath, and Cascade ranges, and the northern Sierra Nevada. This long-lived conifer is surpassed in height only by the redwood (*Sequoia sempervirens*); the tallest known is 326 feet. Needles are spirally arranged around the branchlets, creating a bottlebrush-like appearance (sometimes appear flat).

ID Tips: needles single, spiraled around stem; bark thick, usually deeply furrowed (generally) with age; seed cones with mouse tail bracts

BIGCONE DOUGLAS-FIR

Pseudotsuga macrocarpa

PINACEAE

Height: to 95′ (+175′)
DBH: to 40″
Needles: 1″-1.75″
Seed Cones: 4″-7″

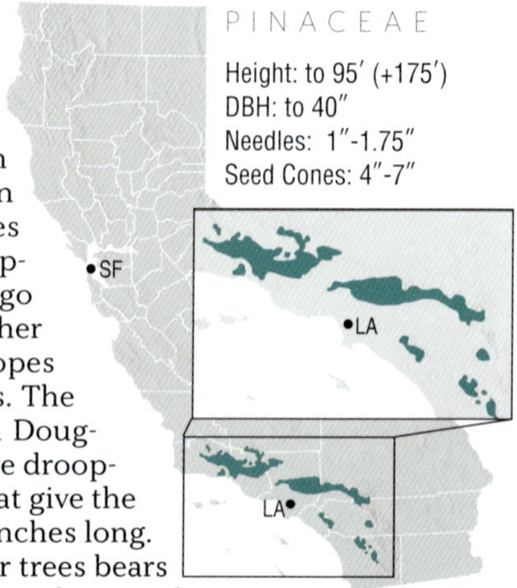

Bigcone Douglas-fir occurs only in southwestern California. It grows in steep ravines and on north-facing slopes in southern California's dry, often chaparral-covered mountains, from San Diego to Santa Barbara counties. It joins other conifers on the forest-covered upper slopes of the Peninsular and Transverse ranges. The seed cones are woodier and larger than Douglas-fir seed cones, and the limbs are more drooping and horizontal. These seed cones, that give the tree its common name, approach eight inches long. The rough, deeply furrowed bark of older trees bears the black scars of past fires. The tallest reaches nearly 175 feet, though specimens are generally much shorter.

ID Tips: needles single; bark thick, deeply furrowed; seed cones exceptionally long; only in Transverse and Peninsular Ranges

Hemlocks

Of the eight species of hemlock worldwide, two live in California, two occur in eastern North America, and the others occur in Asia from the Himalayas to Japan. The seed cones have soft, blunt scales and mature in a single season. The leader, at the top of the tree, droops.

Tsuga mertensiana in the Russian Wilderness, Siskiyou County

WESTERN HEMLOCK

Tsuga heterophylla

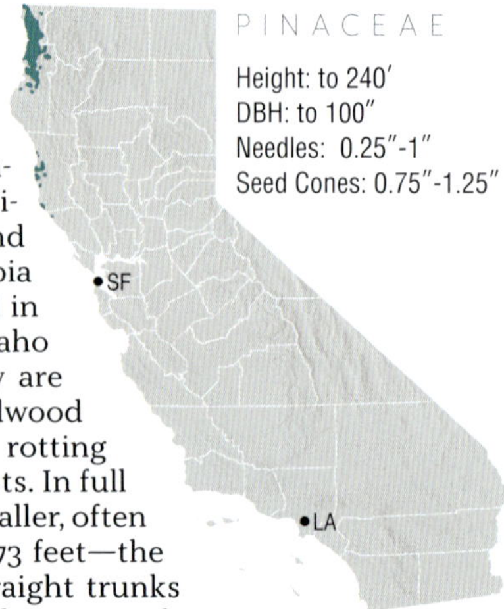

PINACEAE

Height: to 240′
DBH: to 100″
Needles: 0.25″-1″
Seed Cones: 0.75″-1.25″

Western hemlock occurs along the Pacific coast from Sonoma County, California, northward through Oregon and Washington into coastal British Columbia and Alaska. It also grows farther inland in the northern Rockies from northern Idaho and western Montana to Alberta. They are understory trees in Douglas-fir and redwood forests, often germinating on downed, rotting logs, eventually forming miniature forests. In full sun, western hemlocks can grow much taller, often exceeding 150 feet, with the tallest at 273 feet—the tallest of all hemlock species. Their straight trunks are covered in reddish-brown bark and support distinct drooping branch tips. Short, flat, blunt needles, often of different lengths on the same branchlet, are dark green and shiny on the upper side with two bright white lines below. The seed cones are small, with soft, rounded scales.

ID Tips: needles short, green on top and white underneath; cones ~1″ long, egg-shaped, and brown when mature; top branches drooping

MOUNTAIN HEMLOCK

Tsuga mertensiana

PINACEAE

Height: to 175′
DBH: to 75″
Needles: 0.25″-1″
Seed Cones: 1″-3″

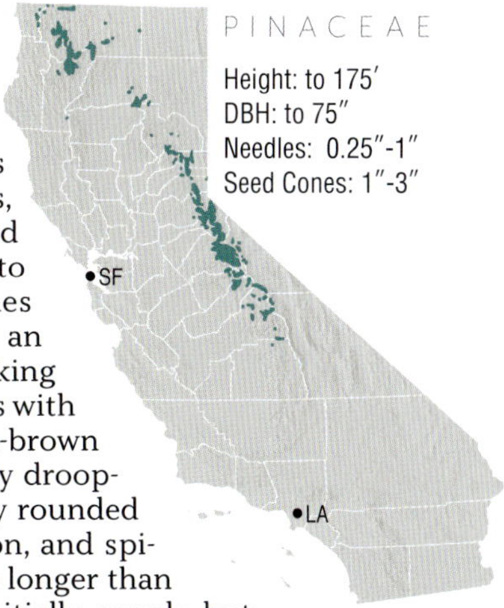

In California, mountain hemlock thrives throughout the Sierra Nevada, Cascades, and Klamath Mountains in protected subalpine areas, ranging northwest into Oregon. The graceful, cascading branches and bluish-green needles of the lend it an elegant, almost sculptural presence, evoking a sense of serene beauty that harmonizes with subalpine landscapes. It has grayish-brown bark and a narrow crown with gracefully drooping branch tips. Its needles have bluntly rounded tips, are round to square in cross-section, and spiral along twigs. The abundant cones are longer than those of western hemlock. They are initially purple but turn a rich brown as the scales peel apart to release the seeds.

ID Tips: needles short, blue-green, growing in all directions around branch; branch tips drooping; high elevations

The Miracle Mile

California is renowned for its conifer diversity, as you have seen while reading this book. However, there is one square mile in California that surpasses all the rest, deep in the Klamath Mountains of northwest California.

The **Miracle Mile of Conifers,** in the Klamath National Forest, is globally significant due to its unparalleled botanical diversity. This remote area harbors 18 native conifer species within a single square mile, making it one of the most conifer-rich regions in the temperate zones. Such a concentration of species is exceptionally rare, as most ecosystems typically host far fewer conifer varieties over much larger areas. This extraordinary assemblage is attributed to the region's unique geology and climate. The **Miracle Mile** serves as a living museum of conifer evolution and a critical refuge for biodiversity, offering vital insights into conservation biology and the effects of climate change on forest ecosystems.

Russian Wilderness, Siskiyou County

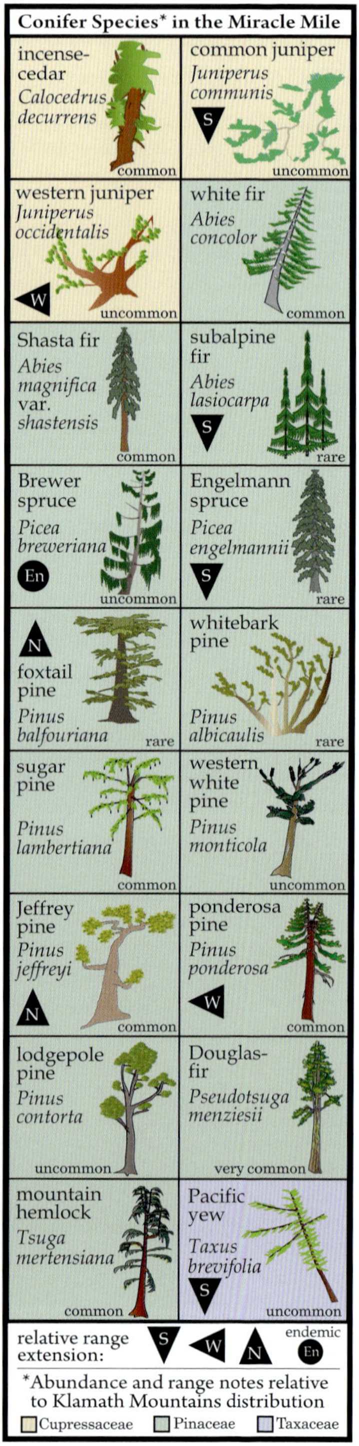

Yew and Nutmeg

The Taxaceae family in North America includes three species in the eastern United States and two in California. In the east, Canada yew (*Taxus canadensis*) thrives in cool, shaded forests, often forming a dense groundcover in moist, nutrient-rich soils in the northern Appalachian Mountains. Both the critically endangered Florida yew (*Taxus floridana*) and Florida torreya (*Torreya taxifolia*) are native to a small region along the Apalachicola River in northern Florida and southern Georgia.

In California, Pacific yew (*Taxus brevifolia*) and the California nutmeg (*Torreya californica*) persist in specific habitats. Each have distinct seed cones that are fleshy and resemble berries (yew) or nuts (nutmeg). The Pacific yew is an understory tree and found in moist, shaded forest environments in northern California. California nutmeg is a lower-elevation inhabitant of the Sierra Nevada foothills and Coast Ranges.

Taxus brevifolia along a stream in northern California, Mendocino County

PACIFIC YEW

Taxus brevifolia

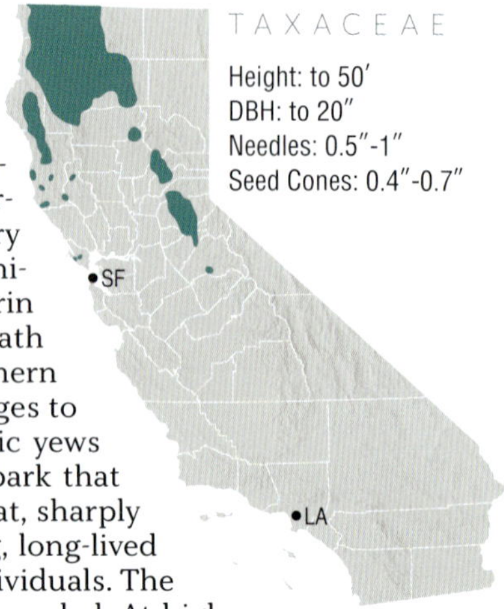

TAXACEAE

Height: to 50′
DBH: to 20″
Needles: 0.5″-1″
Seed Cones: 0.4″-0.7″

Of the eight yew species worldwide, Pacific yew is the only one native to California. It grows in the sheltered understory of Douglas-fir, redwood, and other conifers in the North Coast Ranges from Marin County northward, and in the Klamath Mountains, Cascade Range, and northern Sierra Nevada. Outside California, it ranges to Alaska and eastward to Montana. Pacific yews have fluted trunks covered in reddish bark that curls and peels, and a broad crown of flat, sharply pointed needles. They are slow-growing, long-lived trees with separate male and female individuals. The female cones are red, fleshy, and single-seeded. At higher elevations in the Klamath Mountains, Pacific yew is often shrub-like with yellow-green needles (see photo below right), whereas at lower elevations it is generally tree-like with darker green needles.

ID Tips: bark thin, reddish-brown, scaly; needles single; fruit small, red, berry-like, containing a single seed

CALIFORNIA NUTMEG

Torreya californica

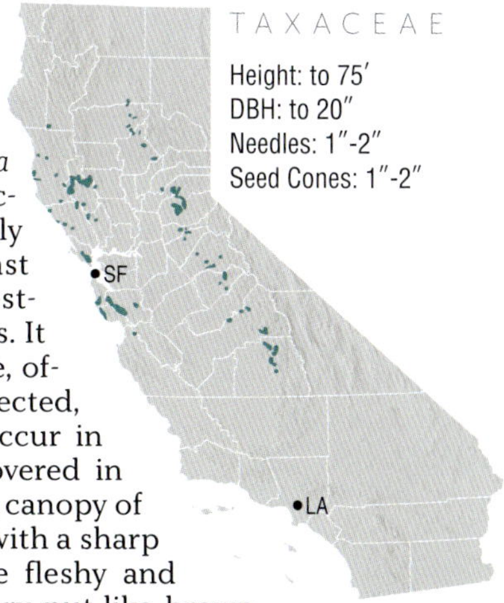

TAXACEAE

Height: to 75′
DBH: to 20″
Needles: 1″-2″
Seed Cones: 1″-2″

California nutmeg is the only *Torreya* in California. The other five species occur in Florida and eastern Asia. Sparsely scattered populations grow in the Coast Ranges, the Cascade Range, and the western Sierra Nevada—never in large groves. It is a slow-growing forest understory tree, often occurring near watercourses in protected, shady canyons but can occasionally occur in chaparral. Straight, tapering trunks covered in furrowed, grayish-brown bark support a canopy of dark green glossy needles, each tipped with a sharp point. Seed cones on female trees are fleshy and slightly larger than an olive. The solitary nut-like brown seed is enclosed by a fleshy outer jacket called an aril, which is green or yellowish-green at maturity, sometimes with maroon stripes.

ID Tips: needles dark green, stiff, tips sharp; fruit with a fleshy covering around large seed, resembling a nutmeg

CONIFEROUS TREES

Flowering Trees

Worldwide, about 60,000 of the 300,000 known plant species grow into trees. Less than one percent of these are conifers. However, over half of the tree species in California are conifers and the rest are flowering trees. Unlike conifers, flowering trees make flowers that, when pollinated, become fruits containing seeds within. They usually have broad leaves, which are deciduous in some species, shedding in autumn and growing back in spring. In California, 17 of our 42 species of flowering trees are evergreen, meaning they retain their leaves year-round. Many flowering trees rely on animal pollinators, such as bees, butterflies, and birds, to transfer pollen between flowers, whereas others are wind-pollinated. In California, flowering trees mostly inhabit coastal areas, valleys, and lower elevation slopes, deserts, and watercourses—often in areas with frequent disturbances like flooding.

The 17 Flowering Tree Families in California (# of Species)

1. Arecaceae (1)
2. Asparagaceae (1)
3. Betulaceae (2)
4. Bignoniaceae (1)
5. Cactaceae (1)
6. Cannabaceae (1)
7. Cornaceae (1)
8. Ericaceae (1)
9. Fabaceae (4)
10. Fagaceae (12)
11. Juglandaceae (2)
12. Lauraceae (1)
13. Oleaceae (2)
14. Platanaceae (1)
15. Rosaceae (1)
16. Salicaceae (7)
17. Sapindaceae (3)

Pictorial Key to the Flowering Plant Families

Palm Family

Arecaceae

California fan palm

pp. 118-119

Asparagus Family

Asparagaceae

Joshua trees

pp. 120-121

Birch Family

Betulaceae

alders & birches

pp. 122-125

Bignonia Family

Bignoniaceae

desert-willow

pp. 126-127

Cactus Family

Cactaceae

giant saguaro

pp. 128-129

Hemp Family

Cannabaceae

netleaf hackberry

pp. 130-131

Dogwood Family

Cornaceae

Pacific dogwood

pp. 132-133

Heath Family

Ericaceae

Pacific madrone

pp. 134-135

Pea Family

Fabaceae

desert ironwood, mesquites, & palo verde

pp. 136-141

Oak Family

Fagaceae

chinquapin, tanoak, & oaks

pp. 142-155

Walnut Family

Juglandaceae

So. & No. California black walnuts

pp. 156-159

Laurel Family

Lauraceae

bay laurel

pp. 160-161

Olive Family

Oleaecae

ashes

pp. 162-165

Sycamore Family

Platanaceae

California sycamore

pp. 166-167

Rose Family

Rosaceae

island ironwood

pp. 168-169

Willow Family

Salicaceae

willows, aspen, & cottonwoods

pp. 170-177

Soapberry Family

Sapindaceae

maples & buckeye

pp. 178-181

Palm Family

The Arecaceae has about 180 genera and 2,600 species, with only one occurring natively in California. Unlike most flowering trees and all conifers, they do not grow annual rings of wood. Yet, palms are trees nonetheless, often with large, single trunks and sizable canopies. Most are frost-tender, from subtropical and tropical parts of the world, with only a few species, such as the Mediterranean fan palm (*Chamaerops humilis*) and Chilean wine palm (*Jubaea chilensis*), occurring in higher latitudes. Palms don't make branches; instead, they have undivided, single trunks, which do not thicken with age. They are significant horticultural plants in southern California, the most common of which are the Mexican fan palm (*Washingtonia robusta*) and the Canary Island date palm (*Phoenix canariensis*). In addition to horticulture, foods and food products come from the delicious fruits of many species: dates (*Phoenix dactylifera*), acai (pronounced awe-saw-EE) (*Euterpe oleracea*), palm oil (*Elaeis guineensis*), and coconut (*Cocos nucifera*).

A *Washingtonia flifera* oasis near Palm Springs, San Bernardino County

CALIFORNIA FAN PALM

Washingtonia filifera

ARECACEAE

Height: to 65′
DBH: to 36″
Leaves: to 13′

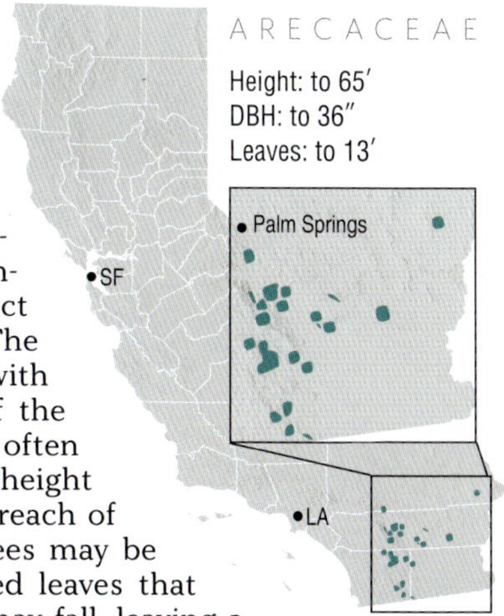

California fan palm grows naturally in isolated groves along streams and around springs and oases in the deserts of the southwestern U.S. and northern Baja California. Their massive, erect trunks may be more than 65 feet tall. The large, fan-shaped leaves are armed with prominent teeth along the margins of the leaf stalks. These protective teeth are often lost when the trees reach over 30 feet, a height at which they would have been out of reach of long-extinct megafauna herbivores. Trees may be skirted with a thatch of dead, recurved leaves that covers the entire trunk, or the leaves may fall, leaving a smooth, bare trunk. Creamy-white flowers and later black, pea-sized fruits grow in large, highly-branched clusters.

ID Tips: trunk tall, columnar, often with a "skirt" of dead leaves around the base; leaves fan-shaped, with thread-like fibers at ends

FLOWERING TREES

Asparagus Family

The Asparagaceae has about 100 genera and 3,000 species. In California this family includes a diverse group of plants, ranging from native species to widely cultivated ornamentals. Native members of this family are primarily found in arid and semi-arid environments, including desert, chaparral, and coastal scrub habitats. Notable genera include *Yucca, Hesperoyucca,* and *Agave*, which are well-adapted to California's dry climate and nutrient poor soils. The family also includes delicate native wildflowers plus several introduced species cultivated for their ornamental value, such as asparagus-fern (*Asparagus aethiopicus*).

Yucca brevifolia in the western Mojave Desert, San Bernardino County

JOSHUA TREE

Yucca brevifolia

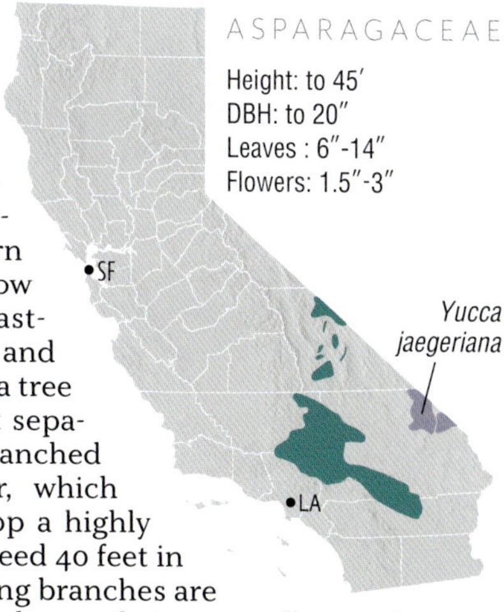

ASPARAGACEAE

Height: to 45′
DBH: to 20″
Leaves : 6″-14″
Flowers: 1.5″-3″

Joshua trees are slow-growing, long-lived, small trees whose range is primarily restricted to the cooler upland portions of the Mojave Desert of southern California. A recent taxonomic split now describes the smaller-statured trees of eastern California, northwestern Arizona, and southwestern Utah as the eastern Joshua tree (*Yucca jaegeriana*), which we do not treat separately here. Joshua trees remain unbranched and columnar until they first flower, which can take 30 years, but they may develop a highly branched canopy in age. The largest exceed 40 feet in height with massive trunks. Their twisting branches are covered in sharply pointed, bayonet-like leaves that eventually turn brown, forming a long-lasting thatch. The trunks of older trees have checkered, grayish-brown bark exposed after the leaves fall. As a keystone species of the Mojave Desert, Joshua trees support an abundance of wildlife that depend on them for food and shelter. Reproduction and that of their sole pollinator, the yucca moth (*Tegeticula synthetica*), are wholly interdependent. Their large, cream-colored flowers go unpollinated if the moth is absent, and the moth must have yucca flowers to complete its life cycle.

ID Tips: unique branching with a single trunk; green-grey leaves in stiff clusters; flowers white, bell-shaped; Mojave Desert

Birch Family

The Betulaceae, which has six genera and 165 species, is represented in California by several ecologically significant tree and shrub species, including alders (*Alnus* spp.) and birches (*Betula* spp.). These plants are primarily found in riparian zones, moist forests, and montane regions, where they contribute to soil stabilization, habitat diversity, and ecological succession. Trees in this family typically have simple, serrated leaves, unisexual flowers in catkins, and lightweight seeds adapted for wind dispersal. Water birch (*Betula occidentalis*), a shrub, is found in wetland or riparian habitats in the north part of the state and along the eastern flanks of the Sierra Nevada.

Four of the world's 25 alder species grow in California, and two are trees. Both are tall and deciduous growing along waterways. Alders are distinguished from birches by features of their fruiting structures, which remain intact in alders but disintegrate in birches.

A champion *Alnus rhombifolia* in the San Gabriel Mountains, Los Angeles County

WHITE ALDER

Alnus rhombifolia

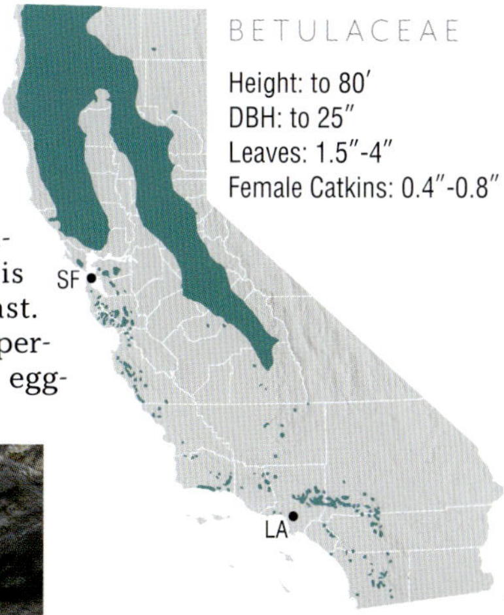

BETULACEAE

Height: to 80′
DBH: to 25″
Leaves: 1.5″-4″
Female Catkins: 0.4″-0.8″

White alder ranges from southern California to Washington, western Idaho, and western Nevada. In California, it occurs throughout much of the state but is less common along the immediate coast. It is never found far from riverbanks or permanent streams. White alder has flat, egg-shaped leaves edged with small teeth.

male catkins

female catkins

ID Tips: bark light gray, smooth when young, scaly with age; leaves glossy above, dull below, dark green, edges finely toothed; riparian

FLOWERING TREES

Forest Pioneers

Alder species in California, such as red alder (*Alnus rubra*) and white alder (*Alnus rhombifolia*), are known as pioneer species because they rapidly colonize disturbed or nutrient-poor habitats. Their roots form symbiotic relationships with nitrogen-fixing bacteria, enriching the soil by converting atmospheric nitrogen into forms usable by plants. This improves soil fertility, making it easier for other plant species to establish. Alders are particularly adept at colonizing areas affected by disturbances such as landslides, floods, wildfires, road cuts, and areas left barren by logging or other human activities. Their ability to thrive in degraded soils, coupled with rapid growth, helps stabilize ecosystems, reinvigorate soils, and prevent erosion.

Alders also provide durable, water-resistant wood that has historically been valued for various applications. Their wood resists decay in wet environments, making it useful for constructing bridges, pilings, and other structures exposed to water. This property also benefits ecosystems, as alder logs and roots in riparian zones create stable habitats for aquatic organisms and help prevent erosion in streambeds. Additionally, alder wood's versatility has made it a favored material for furniture, cabinetry, and tools, further highlighting the ecological and practical importance of these trees.

Alnus rhombifolia pioneering in a stream canyon, Los Angeles County

RED ALDER

Alnus rubra

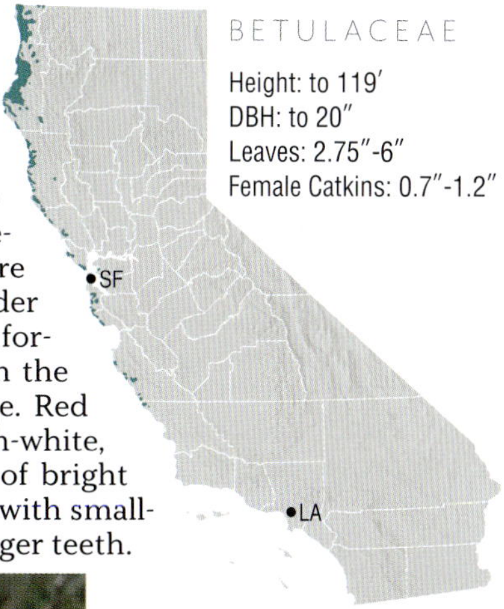

BETULACEAE

Height: to 119′
DBH: to 20″
Leaves: 2.75″-6″
Female Catkins: 0.7″-1.2″

Red alder grows almost entirely near the coast, from Monterey Bay to southern Alaska. Preferring habitats with frequent disturbance like flooding, they are primarily a tree of watercourses. Red alder is becoming more common in upland forests, sometimes forming pure stands in the wetter, more northerly part of its range. Red alders are tall, slender trees with grayish-white, mottled trunks and open, airy crowns of bright green foliage. The leaf edges curl under, with smaller teeth occurring along the edges of larger teeth.

ID Tips: leaf edges coarsely toothed, curling under, dark green above, undersides greenish; coastal

Bignonia Family

The Bignoniaceae is a globally distributed family of flowering plants primarily found in tropical and subtropical regions, with some species extending into temperate zones. It comprises approximately 800 species in about 82 genera. These are characterized by woody vines, shrubs, and trees, with striking, tubular flowers adapted for pollination by birds, bees, and bats. Notable members include the jacaranda (*Jacaranda mimosifolia*), renowned for its vibrant purple blooms and widespread use as an ornamental tree; the trumpet vine (*Campsis radicans*), a vigorous climber native to North America; and the desert-willow (*Chilopsis linearis*), native to arid regions of the southwestern United States and Mexico.

Chilopsis linearis on a dune system in the Mojave Desert, San Bernardino County

DESERT-WILLOW

Chilopsis linearis

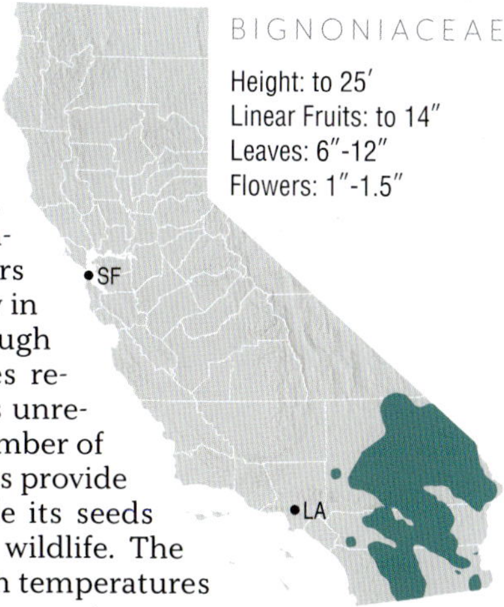

BIGNONIACEAE

Height: to 25′
Linear Fruits: to 14″
Leaves: 6″-12″
Flowers: 1″-1.5″

Desert-willow grows in ephemeral desert washes and along riverbanks and sandy streams throughout the southwestern U.S. and northern Mexico. Flowers form on long, slender branches and vary in color from pink, lavender, or white. Though the linear leaves and weeping branches resemble those of a willow, this species is unrelated to willows. Rather, it is the sole member of the Bignoniaceae in California. Its flowers provide nectar for a variety of pollinators, while its seeds and foliage offer food and shelter for wildlife. The tree's ability to tolerate drought and high temperatures makes it a critical component of desert ecosystems.

ID Tips: open crown; leaves narrow, willow-like; flowers trumpet-shaped, in shades of pink, lavender, or white

Cactus Family

The Cactaceae is significant in California's arid and semi-arid regions. Worldwide, there are approximately 120 genera and 1700 species. These plants are well-adapted to California's deserts. Prominent species include the beavertail cacti (*Opuntia* spp.), chollas (*Cylindropuntia* spp.), and barrel cacti (*Ferocactus* spp.). With their vibrant flowers, cacti also support pollinators such as bees, moths, and bats

Though California's deserts host a diversity of cacti, the saguaro (*Carnegiea gigantea*) has a native range concentrated in the Sonoran Desert of Arizona and northern Mexico. It is one of only a few tree-like members of the Cactaceae.

Carnegiea gigantea

GIANT SAGUARO

Carnegiea gigantea

CACTACEAE

Height: to 50'
DBH: to 30"
Flowers: 3.5"-5"

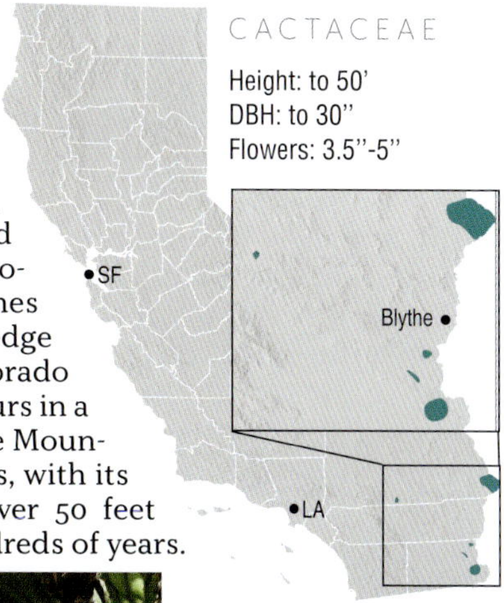

The saguaro is the largest cactus in the U.S., the state flower of Arizona, and a keystone species providing food and shelter for many other species in the Sonoran Desert. The saguaro barely reaches into California on the northwestern edge of its range. It ventures west of the Colorado River only a short distance and only occurs in a few isolated populations in the Whipple Mountains. This slow-growing, massive cactus, with its candelabra-like branches, can reach over 50 feet tall, weigh several tons, and live for hundreds of years.

ID Tips: tallest cactus in U.S.; tree-like, columnar, developing branches or "arms" that bend upward

FLOWERING TREES

Hemp Family

The Cannabaceae includes economically important plants like hemp (*Cannabis sativa*) and hops (*Humulus lupulus*). There are a variety of trees and shrubs worldwide including about 170 species in 11 genera. In North America, this family includes the genus *Celtis*, or the hackberries. Seven of the world's approximately 70 hackberries are native to the U.S., but only one grows natively in California.

Distinct ridged and corky bark of *Celtis reticulata*

NETLEAF HACKBERRY

Celtis reticulata

CANNABACEAE

Height: to 30′
DBH: to 14″
Leaves: 0.75″-3.25″
Flowers: 0.1″ wide

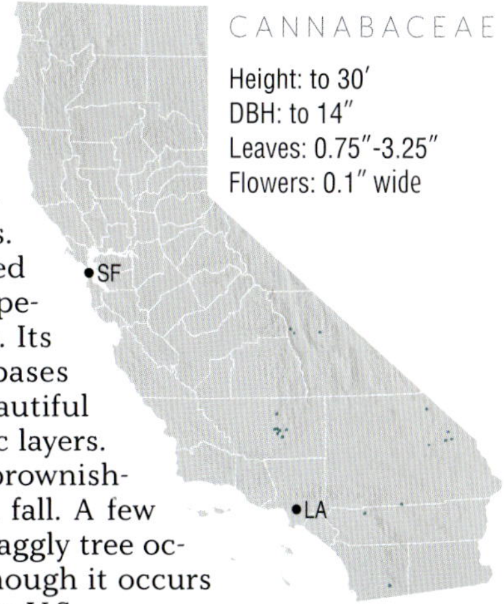

This deciduous tree thrives in arid and semi-arid regions, often inhabiting canyons, rocky slopes, and riparian areas. Recognizable by its rough, net-veined leaves and small, edible drupes, the species has an uncanny drought tolerance. Its leaves are deciduous with asymmetric bases and have a sand-papery feel. The beautiful warty bark peels and folds in concentric layers. Inconspicuous flowers develop into brownish-purple, pea-sized fruits that mature in fall. A few scattered stands of this often small, scraggly tree occur in the southern part of the state, though it occurs widely in other parts of the southwestern U.S.

ID Tips: bark grayish, with noticeable ridges on trunk; leaves alternate, undersides with "net-like" vein pattern

Dogwood Family

The Cornaceae is represented worldwide by about 85 species in only two genera of shrubs and small trees. Of the 16 species of dogwood in the U.S., five grow in California, and one of these is a tree, the Pacific dogwood (*Cornus nuttallii*). Other members of the family in California are shrubs including the red-osier dogwood (*C. sericea*), brown dogwood (*C. glabrata*), blackfruit dogwood (*C. sessilis*), and the rare subshrub western bunchberry (*C. unalaschkensis*). These species support pollinators and birds, as well as being valued for their ornamental beauty and erosion control in restoration projects.

Prolific spring flowers of *Cornus nuttallii*

PACIFIC DOGWOOD

Cornus nuttallii

CORNACEAE

Height: to 75′
DBH: to 12″
Leaves: 2″-5″
Flowers (bracts): to 2.5″

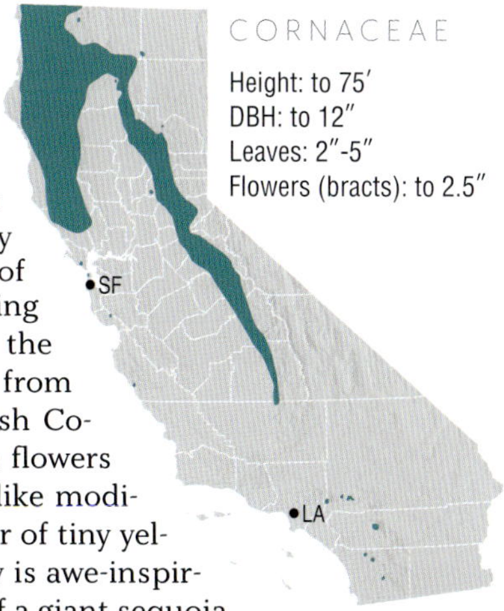

The Pacific dogwood is a small, yet striking deciduous tree found in moist forested areas It is recognized by showy white bracts that surround its clusters of small flowers that usually bloom in spring and occasionally in fall. They grow in the shady understory of conifer forests from southern California northward to British Columbia and eastward to Idaho. It often flowers when leafless, with bright white, petal-like modified leaves that surround a dense cluster of tiny yellowish flowers. This early spring display is awe-inspiring when backed by the reddish trunk of a giant sequoia.

ID Tips: deciduous; leaves opposite, oval-shaped; petal-like bracts large, showy, white; berries bright red, maturing in summer

Heath Family

The Ericaceae is a diverse group of flowering plants found worldwide, particularly in temperate and subarctic regions, represented by over 4,000 species in about 120 genera. This family encompasses a wide range of species, from evergreen shrubs to small trees, including familiar plants like blueberries and cranberries (*Vaccinium* spp.), rhododendrons (*Rhododendron* spp.), and heathers (*Calluna vulgaris*, *Erica* spp.). In North America, the family is well-represented by species thriving in acidic, nutrient-poor soils, such as the mountain laurel (*Kalmia latifolia*), azaleas (*Rhododendron* spp.), and various wild blueberries. In California, the family is notable for its chaparral and montane species, including manzanitas (*Arctostaphylos* spp.), western azalea (*Rhododendron occidentale*), and the only native species in the family that consistently grows into a tree: the Pacific madrone (*Arbutus menziesii*).

A massive *Arbutus menziesii*, lost in the 2016 Soberanes Fire, Monterey County

PACIFIC MADRONE

Arbutus menziesii

ERICACEAE

Height: to 80′
DBH: to 30″ (+50″)
Leaves: 3″-6″
Flowers: 0.3″ (in clusters)

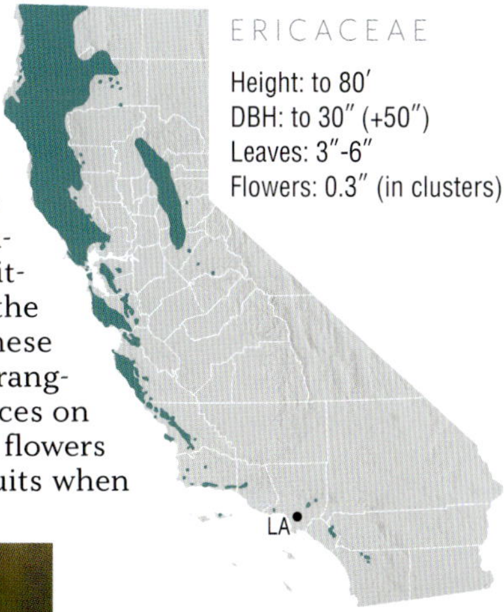

Of the 20 species of *Arbutus*, only one is native to California. Pacific Madrones occur in patchy stands in coastal mountains from Baja California, Mexico to British Columbia, Canada and inland in the Cascade Range and Sierra Nevada. These beautiful evergreen trees have smooth, orangish-red bark and glossy upper leaf surfaces on two-toned leaves. White, lantern-shaped flowers emerge in spring, forming red, warty fruits when pollinated.

ID Tips: bark smooth and peeling, reddish-brown; leaves evergreen, glossy above, oval-shaped; flowers bell-shaped, white; berries red in fall

FLOWERING TREES

Pea Family

The Fabaceae, also legume or bean family, is one of the largest plant families globally, with over 20,000 species in about 750 genera. There are more trees in the Fabaceae than any other family. Members in this family are ecologically important and, like alders, have the ability to fix atmospheric nitrogen through symbiotic relationships with soil bacteria. In California, the legume family includes four native trees in three genera that grow in arid regions.

The genus *Olneya* is represented by a single species, the desert ironwood. This tree is native to the Sonoran Desert in North America.

California's two species of mesquites (*Neltuma odorata* and *Strombocarpa pubescens*) were formerly classified in the genus *Prosopis*, which was recently split apart. Mesquites are spiny trees with durable, dense wood and deep root systems that thrive in dry habitats. Their leathery legumes are edible and rich in sugar.

Parkinsonia spp. (palo verdes) are thorny, drought-deciduous trees and shrubs native to arid regions of the Americas and Africa. In California, this genus is represented by the blue palo verde (*P. florida*) and several shrubs including foothill palo verde (*P. microphylla*). These trees are characterized by their green, photosynthetic bark.

Olneya tesota in the Sonoran Desert

DESERT IRONWOOD

Olneya tesota

FABACEAE

Height: to 60′
DBH: to 24″
Leaves: 0.25″-1″
Seedpods: 2″-3″

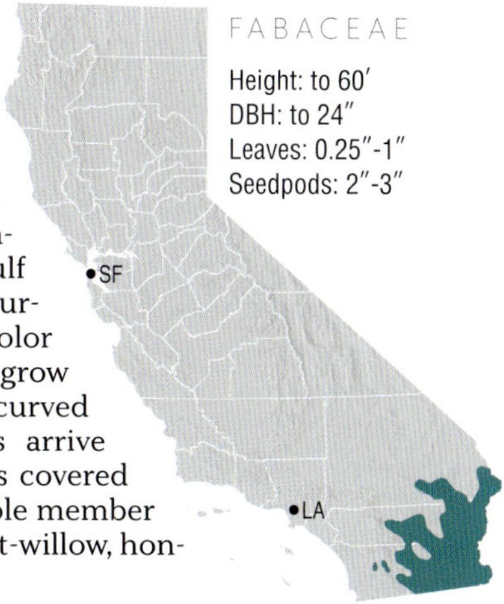

Ironwood grows in washes and gravelly flats in southeastern California's Sonoran Desert, ranging eastward to central Arizona and southward to the Gulf of California. Thin, rough, scaly bark surrounds exceedingly dense wood, the color of dark chocolate. Feather-like leaves grow on slender shoots protected by pairs of curved spines. Pinkish-white, pea-like flowers arrive in summer and yield leathery seedpods covered with fine hair. Desert ironwood is the sole member of its genus and often occurs with desert-willow, honey mesquite, and blue palo verde.

ID Tips: bark rough, scaly; branches with paired thorns; leaves compound, grayish-green; flowers pinkish-white

HONEY MESQUITE

Neltuma odorata

FABACEAE

Height: to 45′
DBH: to 15″
Leaflets: 2″-7″
Seedpods: 2″-8″

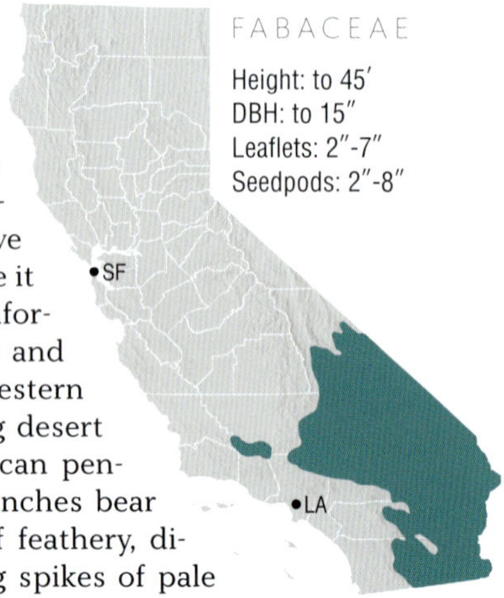

Honey mesquite is a shrub or small tree that grows in much of the southwestern U.S. and Mexico. Outside of its native range, it grows in many countries where it is considered an invasive weed. In California, it ranges throughout the Sonoran and Mojave deserts and northward to the western San Joaquin Valley. It is frequent along desert washes and on sand dunes. Its roots can penetrate to great depths. The limber branches bear paired, straight spines at the bases of feathery, divided leaves. Honey mesquite has long spikes of pale yellowish-brown flowers and abundant, edible seedpods.

ID Tips: tree or shrub; thorns paired, straight; leaves twice-compound, feathery; flowers small, yellow; seedpods yellowish-brown

SCREWBEAN MESQUITE

Strombocarpa pubescens

FABACEAE

Height: to 25′
DBH: to 10″
Leaflets: 1″-2″
Seedpods: 2″-3″

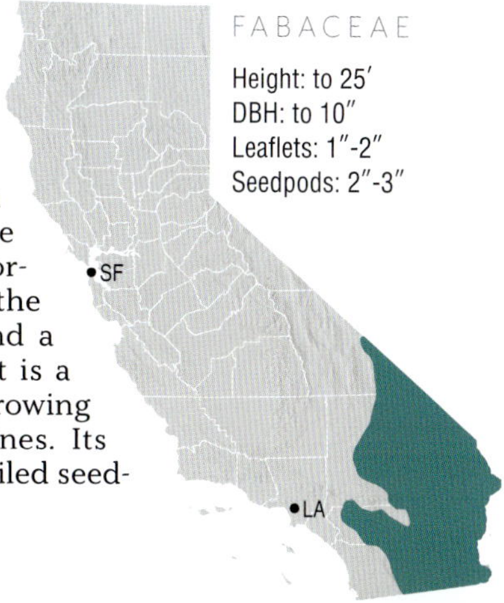

Screwbean mesquite grows in gravelly washes and desert ravines, especially in areas with saline soils, throughout the Sonoran and Mojave deserts of California. Its range extends eastward across the southwestern U.S. to western Texas and a short distance into northern Mexico. It is a small tree with twice-divided leaves growing along branches armed with short spines. Its most distinctive feature is the tightly coiled seedpod.

ID Tips: tree or shrub; bark shaggy, shedding; flowers catkin-like, yellow; seed pods tightly spiraled

Stem Photosynthesis

Photosynthesis is the process by which green plants, algae, and some bacteria convert sunlight, carbon dioxide, and water into glucose (sugar) and oxygen. This process is vital because it serves as the foundation for food chains that sustain life on Earth. Additionally, photosynthesis plays a crucial role in maintaining atmospheric oxygen levels and reducing carbon dioxide—helping regulate the Earth's climate and supporting the survival of countless species, including humans. Photosynthesis in land plants mostly occurs in leaves.

Although leaves are the typical photosynthetic organs for plants, many woody desert shrubs and subshrubs utilize green stems as important, or even primary, photosynthetic organs. These species commonly have small, ephemeral leaves that are present for just a few days or weeks in spring. This functionally leafless trait is common in members of the legume and sunflower families and in stem succulents like giant saguaro.

Photosynthetic green stems are generally better at water conservation than leaves and therefore are good choices for positive net photosynthesis during drought months. Although one might assume that a green stem should lose relatively little water because it is essentially leafless, some desert plant stems have moderate rates of water loss through transpiration. Consequently, many of these shrubs, with their large total stem surface area, also have relatively high requirements for water. This may explain why many of these stem-photosynthetic species in our deserts either inhabit washes and other drainage channels or have deep roots to tap ephemeral pools of water.

Stem photosynthesis (from L to R) blue palo verde, quaking aspen, Pacific madrone

BLUE PALO VERDE

Parkinsonia florida

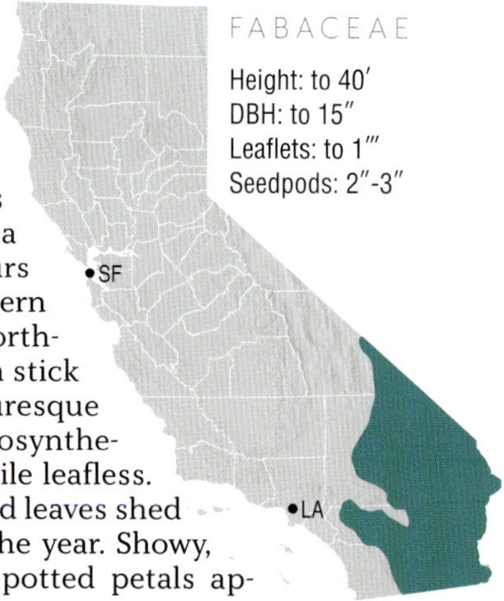

FABACEAE

Height: to 40′
DBH: to 15″
Leaflets: to 1‴
Seedpods: 2″-3″

Palo verdes are a group of about a dozen spiny, golden-flowered shrubs and trees native to hot deserts of Africa and the Americas. Blue palo verde occurs in the Sonoran Desert from southeastern California into central Arizona and northwestern Mexico. Palo verde means green stick in Spanish, alluding to how these picturesque small trees perform much of their photosynthesis with their greenish-yellow stems while leafless. Like other palo verdes, the finely textured leaves shed their tiny leaflets during dry times of the year. Showy, fragrant flowers with golden, orange-spotted petals appear in late spring and early summer.

ID Tips: bark on trunk and stems blue-green; leaves small, sparse (often absent); flowers golden, often appearing when tree is mostly leafless

FLOWERING TREES

Oak Family

The Fagaceae, commonly known as the oak or beech family, includes notable genera such as oaks (*Quercus*), beeches (*Fagus*), and chestnuts (*Castanea*). Worldwide, there are over 900 species in the family in eight genera. In North America, members of the Fagaceae are vital components of various ecosystems, ranging from deciduous forests to savannas. There are three genera in California.

Chrysolepis comprises two species, *C. sempervirens* (bush chinquapin) and *C. chrysophylla* (giant chinquapin), of the Pacific states. These evergreens are known for their spiny burs that encase edible nuts.

Tanoak (*Notholithocarpus densiflorus*) is unique to the Pacific Coast and thrives in California's coastal and mixed-evergreen forests. Although resembling an oak, tanoak is distinct in its evolutionary lineage and several physical characteristics such as non-hanging male flowers.

Oaks are characteristic California trees, and their woodlands cover 10% of the state. Of the 600 oak species, 21 are native to California, and 10 regularly grow into trees. The others are mostly chaparral shrubs. Most oaks in California are restricted to the western part of the state, northern Baja California, and southwestern Oregon. California's oaks, like the canyon live oak (*Quercus chrysolepis*) and the valley oak (*Q. lobata*), are some of the largest in the world.

Quercus tomentella on Santa Rosa Island, Santa Barbara County

GIANT CHINQUAPIN

Chrysolepis chrysophylla

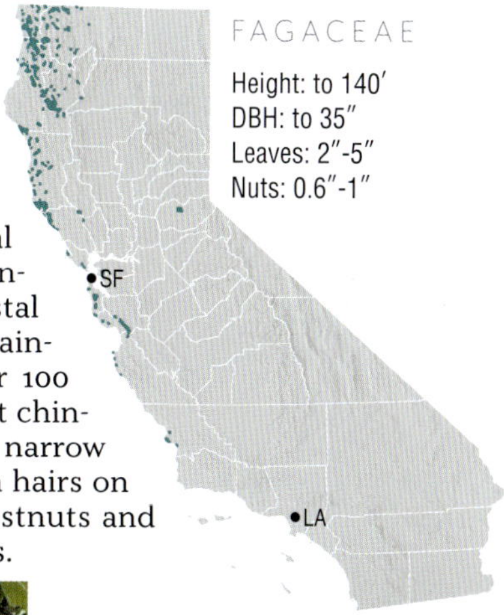

FAGACEAE

Height: to 140′
DBH: to 35″
Leaves: 2″-5″
Nuts: 0.6″-1″

There are two *Chrysolepis* species, both in California, and one grows into a tree. Giant chinquapin ranges from central California to southern Washington, mainly as an occasional component of coastal forests. In favorable conditions of high rainfall, this evergreen tree can reach over 100 feet tall but is often much smaller. Giant chinquapins have fluted trunks and long, narrow leathery leaves covered with fine golden hairs on the undersides. They are related to chestnuts and oaks and make edible nuts in spiny cups.

ID Tips: bark thick, furrowed; leaves leathery, dark green above, undersides golden-yellow; fruits spiny, bur-like

FLOWERING TREES

TANOAK

Notholithocarpus densiflorus

FAGACEAE

Height: to 130′
DBH: to 75″
Leaves: 3″-7″
Acorns: 0.75″-1.25″

Tanoak is widespread in woodlands and evergreen forests in coastal mountains from Ventura County to southern Oregon and in the interior from Fresno County northward. Tanoak is related to oaks (*Quercus* spp.) and makes edible, acorn-like nuts, but its clusters of male flowers are stiff and upright (they are pendent in the oaks). Its veiny evergreen leaves are thick and leathery and can stay on the tree for several years. They emerge covered in dense golden hairs that soon wear off. This species is significantly affected by Sudden Oak Death, a disease caused by a water mold (*Phytophthora ramorum*) likely from Asia. First discovered in California in the mid-1990s, the pathogen has created devastating tanoak die-offs throughout its range. In the late 1800s, the California leather industry used the bark for tanning.

ID Tips: evergreen; bark thin, with flattened ridges; leaves leathery, edges smooth or slightly toothed; spiky acorn cup scales

COAST LIVE OAK

Quercus agrifolia

FAGACEAE

Height: to 100′
DBH: to 150″
Leaves: 0.75″-2.5″
Acorns: 0.75″-1.5″

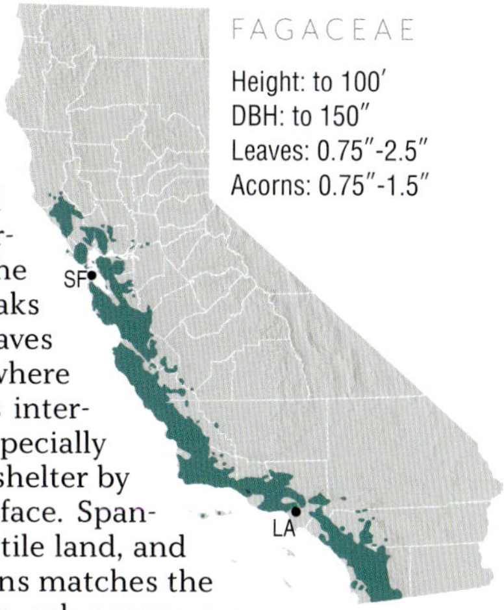

Coast live oak is an evergreen tree that can live for several hundred years and forms the backbone of coastal California woodlands. A California icon, it is the basis for city names like Thousand Oaks and Oakland. Its leathery, cupped leaves have tufts of hair on the undersides where the main vein and the secondary veins intersect. These pockets of hair, which are especially present on younger leaves, are used as shelter by mites that graze fungi on the leaf surface. Spanish settlers associated this tree with fertile land, and the placement of the Franciscan missions matches the coast live oak's native range. Coast live oak acorns mature in one year, unlike the other three species of California red oaks, whose acorns require two years to ripen.

ID Tips: bark gray, furrowed; leaves evergreen, leathery, dark green, undersides glossy with "hairy armpits" (pic above), edges spiny, curling under

CANYON LIVE OAK

Quercus chrysolepis

FAGACEAE

Height: to 100′
DBH: to 150″
Leaves: 1″-2.5″
Acorns: 0.75″-2″

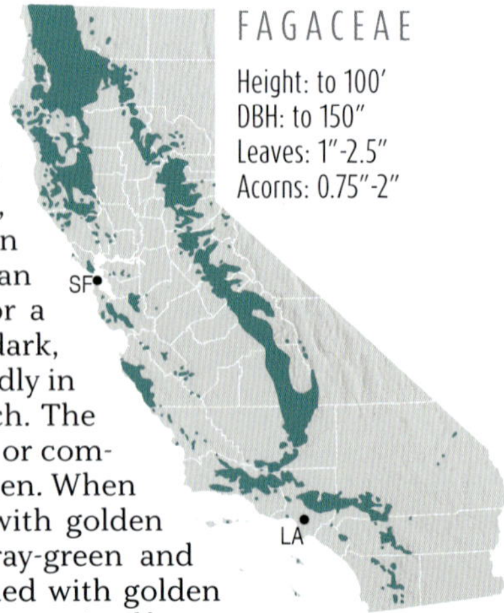

Canyon live oak or maul oak is the most widely distributed oak in California, ranging from Baja California to southern Oregon and eastward into Arizona. It can be a small shrub on exposed ridges or a massive tree in deep canyons. It has a dark, evergreen canopy of leaves that vary wildly in shape and size, often on a single branch. The leaf edges are irregularly spine-toothed or completely smooth and everything in between. When young, the leaf underside is covered with golden hairs but eventually becomes pale gray-green and hairless. The acorn cups, which are lined with golden felt, are generally much wider than the acorn itself.

ID Tips: leaves evergreen, leathery, often spiny-toothed, dark green above, golden-fuzzy or dull gray-green beneath; acorn cup golden-hairy

World's Largest Oaks

Oak trees are among the most iconic and long-lived species in the world, known for their grandeur and ecological importance. Some of the largest oak trees stand as natural monuments, showcasing the incredible diversity and adaptability of the oak family. From ancient forests in Europe to towering specimens in North America, these trees captivate with their sheer size, age, and resilience.

In the southeastern United States, the Angel Oak, a southern live oak (*Quercus virginiana*) in South Carolina, is one of the largest and oldest oaks in the country, with a canopy spreading over 17,000 square feet and an estimated age of 400 to 500 years. Its sprawling limbs create an enchanting landscape, drawing visitors from around the globe. In Europe, the Quercia Vallonea of Tricase in Italy is another giant, with an estimated age of 700-900 years.

The canyon live oaks of California are reaching legendary status. The current record holder by American Forest standards is in the San Bernardino Mountains. There are similarly large specimens spread around the state, including the San Gabriel Oak in Los Angeles County, the Smither's Oak in Mariposa County, and the Iaqua Oak in Humboldt County. One of these four is most likely the largest oak, by volume, on Earth.

Smither's Oak

Wildlands Conservancy Oak

Iaqua Oak

San Gabriel Oak

BLUE OAK

Quercus douglasii

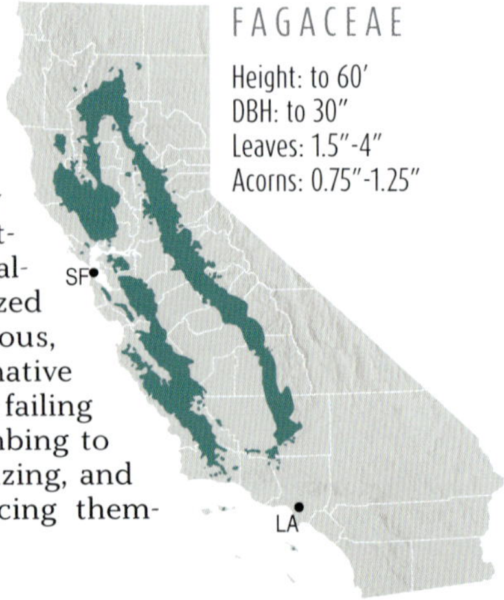

FAGACEAE

Height: to 60′
DBH: to 30″
Leaves: 1.5″-4″
Acorns: 0.75″-1.25″

Blue oak is California's most drought-tolerant tree oak and grows natively only in the state. It thrives in the hot, dry foothills surrounding the Great Central Valley. A mature blue oak is a medium-sized tree with a rounded canopy of deciduous, dull bluish-green leaves. In much of its native range, blue oak reproduction has been failing for nearly a century. Old trees, succumbing to drought, weeds, intensive livestock grazing, and development, are not naturally replacing themselves.

ID Tips: deciduous; crown rounded; bark checkered with small scales, light gray; leaves shallowly lobed or smooth-margined, bluish-green

ENGELMANN OAK

Quercus engelmannii

FAGACEAE

Height: to 50′
DBH: to 25″
Leaves: 1.25″-2.25″
Acorns: 0.5″-1″

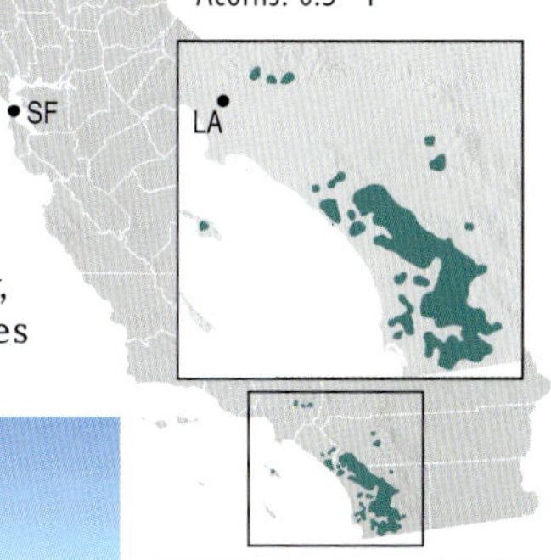

Engelmann or mesa oak is a semi-evergreen species growing only in savannas and foothill woodlands from northern Baja California to Pasadena. Its range is smaller than that of most California oaks, and suburban sprawl has displaced it in many areas. Thick trunks covered in furrowed, grayish-brown bark support a low, broad canopy of leathery, bluish-gray leaves with smooth, wavy edges.

ID Tips: semi-evergreen; crown rounded; bark thick, furrowed, light gray-brown; leaves leathery, oblong to oval, blue-green, edges slightly wavy

FLOWERING TREES

OREGON OAK

Quercus garryana

FAGACEAE

Height: to 80′
DBH: to 36″
Leaves: 3″-6″
Acorns: 0.75″-1″

Oregon oak is the only oak in the state to extend far beyond California's borders. It grows from central California northward into Oregon, Washington, and British Columbia. Two shrubby varieties, *breweri* and *semota*, are scattered in the Klamath Mountains and Sierra Nevada foothills, respectively. The Oregon oak's stout trunk is covered in whitish-gray, checkered bark and supports broad limbs with deciduous, glossy, deeply lobed leaves.

ID Tips: deciduous; crown broad, spreading; leaves with rounded lobes, dark green; acorn cups shallow, wider than deep

FLOWERING TREES

CALIFORNIA BLACK OAK

Quercus kelloggii

FAGACEAE

Height: to 80′
DBH: to 42″
Leaves: 4″-9″
Acorns: 1″-1.5″

Of California's four native species of red oak (see *Q. wislizeni* treatment), only California black oak has deeply lobed deciduous leaves. All four species form natural hybrids, the most common of which is the oracle oak (*Q.* × *morehus*), a cross between California black oak and interior live oak. California black oak grows in valleys, on mountain slopes, and along open ridges from the Mexican border to central Oregon. It is more cold-tolerant than most other California oaks and can be found growing among montane conifers. Graceful trunks covered in black bark support a broad canopy of deeply lobed leaves, each lobe terminating in a long bristle.

ID Tips: deciduous; bark dark, becoming platy; leaves deeply lobed with bristle-tips; acorns mature in second year

VALLEY OAK

Quercus lobata

FAGACEAE

Height: to 100′
DBH: to 100″
Leaves: 2″-4″
Acorns: 0.75″-1.5″

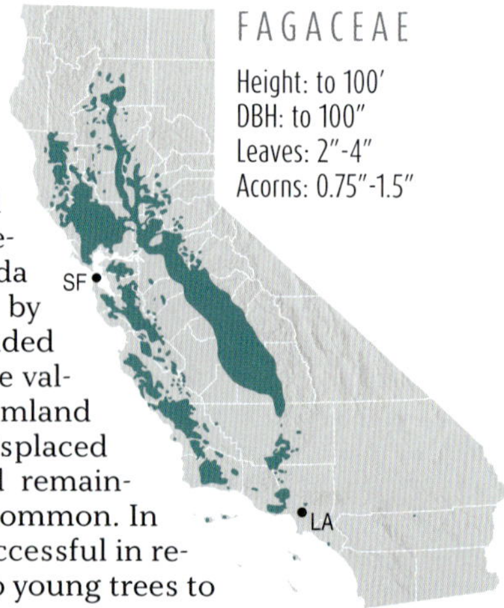

Valley oak is among the world's largest oaks. It grows only in the Great Central Valley and surrounding ranges, especially the Coast Ranges, and Sierra Nevada foothills. Its massive trunk, covered by deeply checkered bark, supports a rounded canopy of deciduous, lobed leaves. Since valley oaks grow on deep, moist loam, farmland development over the last 200 years displaced many trees throughout its range, and remaining stands of mature trees are now uncommon. In many areas, valley oaks have been unsuccessful in reproduction; as old trees die, there are no young trees to replace them.

ID Tips: deciduous; bark with "checkerboard" pattern; leaves deeply lobed, lobes rounded; acorns long, pointed

SHREVE OAK

Quercus parvula var. *shrevei*

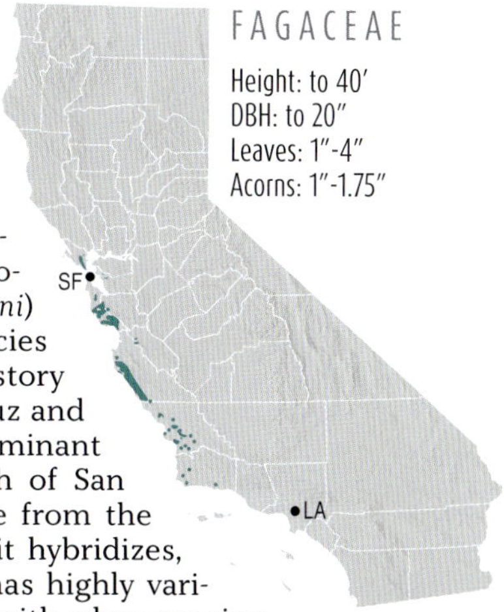

FAGACEAE

Height: to 40′
DBH: to 20″
Leaves: 1″-4″
Acorns: 1″-1.75″

Shreve oak is the tree form variety of *Quercus parvula*, a species named initially for its shrubby form on Santa Cruz Island in 1880. It was then combined taxonomically with interior live oak (*Q. wislizeni*) but was recognized as a separate species a century later. Shreve oak is an understory tree of the redwood forests in Santa Cruz and Monterey counties and is one of the dominant trees in mixed-evergreen forests. North of San Francisco, it is nearly indistinguishable from the two other red oak species with which it hybridizes, coast live oak and interior live oak. It has highly variable, flat, lance-shaped evergreen leaves with edges ranging from spiny to smooth and undersides that are yellowish-green.

ID Tips: evergreen; leaves small, dark green, edges smooth or slightly spiny, undersides lacking "hairy armpits" of coast live oak

ISLAND OAK

Quercus tomentella

FAGACEAE

Height: to 65′
DBH: to 20″
Leaves: 2.75″-4″
Acorns: 1″

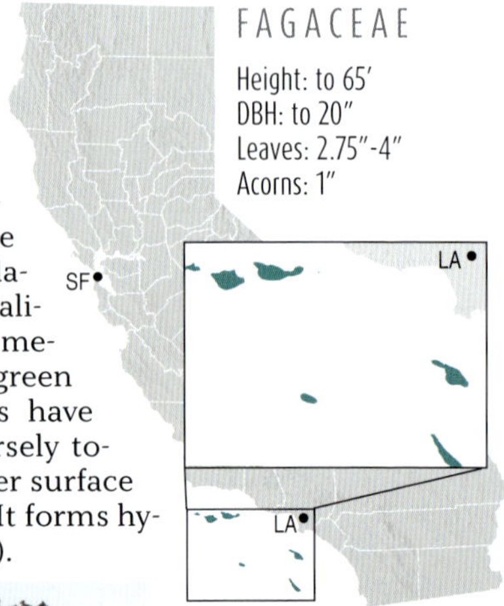

Though it is widely grown horticulturally, island oak is the rarest of California's oaks, growing natively only on five California Channel Islands and Guadalupe Island off the west coast of Baja California, Mexico. It is a smooth-barked, medium-sized tree with a rounded, dark green canopy. The leathery, evergreen leaves have pleated parallel side veins and are sparsely tomentose or becoming shiny on the upper surface and covered with woolly hairs beneath. It forms hybrids with canyon live oak (*Q. chrysolepis*).

ID Tips: evergreen; medium-statured; leaves large, finely hairy above, densely hairy below

INTERIOR LIVE OAK

Quercus wislizeni

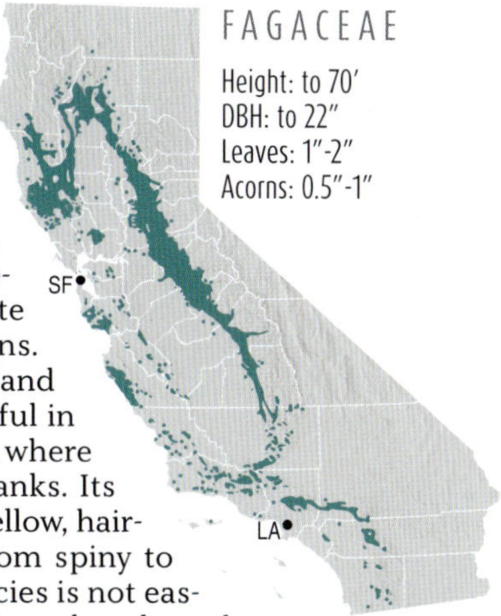

FAGACEAE

Height: to 70′
DBH: to 22″
Leaves: 1″-2″
Acorns: 0.5″-1″

California has four species of red oak (*Quercus* section *Lobatae*, series *Agrifoliae*). Three of these, coast live oak (*Q. agrifolia*), California black oak (*Q. kelloggii*), and interior live oak, are widespread in the state and have broadly overlapping distributions. The fourth, *Q. parvula*, is a coastal and island species. Interior live oak is most successful in the western foothills of the Sierra Nevada, where it grows with blue oak on dry alluvial banks. Its leaves are shiny and flat, with greenish-yellow, hairless undersides and edges that range from spiny to smooth. North of San Francisco, this species is not easily distinguished from Shreve and coast live oaks, where the three species blend in a swarm of hybridization resulting in widely varying individuals.

ID Tips: evergreen tree or shrub; crown broad; leaves glossy, dark green above, light green below, edges smooth or with spiny teeth

FLOWERING TREES

Walnut Family

The Juglandaceae is a globally distributed family of flowering trees with about 50 species in 9 genera. These trees are renowned for their ecological, economic, and cultural significance, particularly for their edible nuts and high-quality timber. Globally, members of the Juglandaceae can be found in temperate and subtropical regions, with species like walnuts (*Juglans* spp.) and hickories (*Carya* spp.) being widely cultivated.

There are 20 species of walnut worldwide and six in the U.S. The two species native to California occur only within the state. California has a long history of walnut farming, especially in the Sacramento Valley where there are many orchards of English walnut (*J. regia*) and hybrids with the Northern California black walnut.

Juglans hindsii

SO. CA. BLACK WALNUT

Juglans californica

JUGLANDACEAE

Height: to 50′
DBH: to 10″
Leaflets: 11-19
Seeds: 1.5″-2.5″

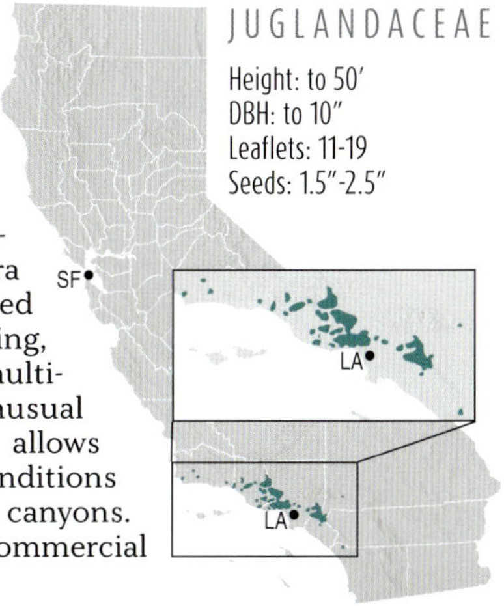

This small, winter-deciduous tree has a spotty distribution mainly in the low mountains and riparian areas in southern California from Orange to Ventura counties. Bright green, pinnately divided leaves form a crown above a low-branching, dark gray trunk. It often grows as a multi-trunked tree or large shrub, which is unusual for walnut species. This growth habit allows it to thrive in the arid and rugged conditions of southern California's foothills and canyons. Though its walnuts are smaller than commercial walnuts, they are equally delicious.

ID Tips: deciduous; multi-trunked tree or shrub; bark thick, deeply furrowed; leaves pinnately compound, with 11-19 lanceolate leaflets

Cultural Significance of California's Native Walnuts

California's two native walnut species, the northern California black walnut (*Juglans hindsii*) and the southern California black walnut (*Juglans californica*), hold deep cultural significance, reflecting their integral role in the region's history, ecology, and identity. These trees, endemic to California, have been intertwined with the lives of Indigenous peoples, settlers, and modern conservationists, becoming symbols of resilience in a rapidly changing landscape.

Indigenous Importance

For First Peoples, pre-1850, California's native walnuts were a vital resource. The hard-shelled nuts provided a nutritious food source rich in fats and protein that can be consumed fresh or stored for later use. Tribes such as the Tongva, Chumash, and others in southern California incorporated walnut seeds into their diets and used the wood for crafting tools, implements, and ceremonial items. The trees were also woven into spiritual practices, embodying respect for nature and the interconnectedness of ecosystems.

Modern Uses

During the 19th and early 20th centuries, settlers recognized the value of California's native walnuts, particularly for their durable wood. The wood's rich, dark tones made it highly desirable for fine furniture, cabinetry, and gunstocks. Despite their smaller size compared to cultivated English walnuts, the nuts from native trees were occasionally harvested and used locally. As agriculture expanded, these native walnuts became crucial in the walnut industry, serving as hardy rootstocks for grafting the commercially popular English walnut (*Juglans regia*).

Northern California Black Walnut
Juglans hindsii

NOR. CA. BLACK WALNUT

Juglans hindsii

JUGLANDACEAE

Height: to 60′
DBH: to 20″
Leaflets: 13-21
Seeds: 1.5″-2.5″

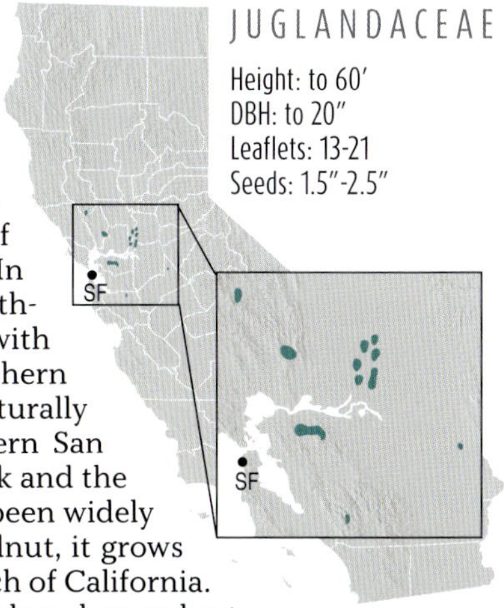

Northern California black walnut is a much larger tree than its southern California cousin. It can achieve heights of 60 feet and can form a massive trunk. In addition to the size difference, the northern species has larger nuts and leaves with tufts of hair on the undersides. Northern California black walnut only occurs naturally in a few populations in the northeastern San Francisco Bay Area around Walnut Creek and the lower Sacramento River. Because it has been widely cultivated as a rootstock for English walnut, it grows outside its native range throughout much of California. Its beautiful purplish-brown wood is sold as claro walnut.

ID Tips: deciduous; bark dark with distinct smell; leaves long, pinnately compound, with 13-21 lanceolate, finely toothed leaflets

FLOWERING TREES

Laurel Family

The Lauraceae is a large and diverse family with over 40 genera and about 2,800 species, distributed primarily in tropical and subtropical regions. Members of the Lauraceae, such as avocados (*Persea americana*) and camphor trees (*Cinnamomum camphora*), are valued for their aromatic leaves, fruits, and timber. In California, the family is represented only by the bay laurel (*Umbellularia californica*), also known as the Oregon myrtle or pepperwood. This native tree thrives in the coastal and inland forests of the state, where it serves as an important ecological and cultural species.

Umbellularia californica can reach immense sizes

BAY LAUREL

Umbellularia californica

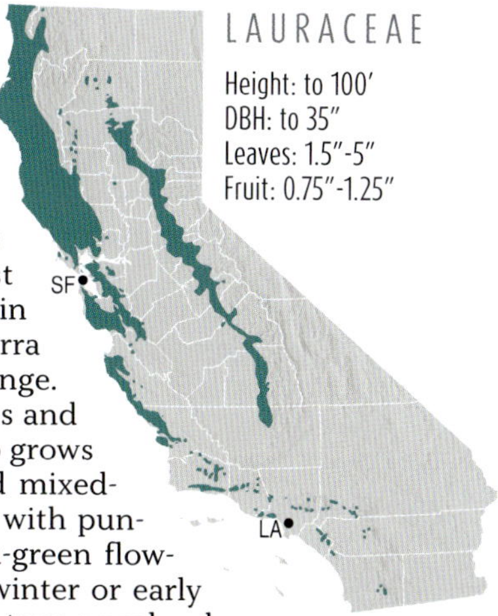

LAURACEAE

Height: to 100′
DBH: to 35″
Leaves: 1.5″-5″
Fruit: 0.75″-1.25″

California bay laurel grows from Baja California to southern Oregon. Near the coast, it occurs from the mountains of southern California through the Coast Ranges to the Klamath Mountains and in the interior along the length of the Sierra Nevada to the foothills of the Cascade Range. It is associated with California sycamores and oaks in shaded riparian habitats but also grows in oak woodlands, redwood forests, and mixed-evergreen forests. It is an evergreen tree with pungently scented leaves. It bears yellowish-green flowers in umbrella-shaped clusters in late winter or early spring, followed by olive-sized fruits that turn purple when ripe. The fine-grained, honey-colored wood, sometimes called pepperwood or Oregon myrtle, is used in furniture, bowl turning, and carving.

ID Tips: broadleaf evergreen tree, can be shrub-like at higher elevations; leaves lance-shaped, leathery, dark green, highly aromatic

FLOWERING TREES

Olive Family

The Oleaceae is globally distributed and that includes around 25 genera and 700 species. Found across temperate and tropical regions, the family encompasses a wide variety of trees, shrubs, and vines, including olive (*Olea europaea*), ashes (*Fraxinus* spp.), jasmines (*Jasminum* spp.), and lilacs (*Syringa* spp.). Globally, species in the Oleaceae are valued for their timber, ornamental beauty, and agricultural importance. The 65 species of ash are deciduous trees or large shrubs from the Northern Hemisphere. They have compound leaves in pairs, unremarkable wind-pollinated flowers, and winged fruits. Five species are native to California, and two grow as trees. These two species are closely related and have been treated as one in many guides. They have scattered occurrences in wet areas from San Diego County to the Oregon border, with blurred boundaries between their distributions.

The champion *Fraxinus latifolia* grows in Del Norte County

OREGON ASH

Fraxinus latifolia

OLEACEAE

Height: to 80′
DBH: to 30″
Leaves: 5-9 leaflets
Fruit: 1″-2″

Oregon ash grows in canyons, on streambanks, and in swampy woodlands from central California to British Columbia. In California, it occurs in the Coast Ranges from the San Francisco Bay region northward through the Klamath Mountains and in the interior throughout the length of the Sierra Nevada. Its leaflets are rounder and with shorter stalks than the more spearhead-shaped leaves of velvet ash. Easily confused with velvet ash where ranges overlap.

ID Tips: deciduous; crown broad, rounded; leaves compound, with 5-9 leaflets; fruits winged (samaras); generally near waterways

VELVET ASH

Fraxinus velutina

OLEACEAE

Height: to 30′
DBH: to 10″
Leaves: 5 or 7 leaflets
Fruits: 0.75″-1.25″

Velvet ash grows along watercourses and in other wet areas from southern California eastward to southwestern Utah, Texas, and northern Mexico. In California, it grows in scattered populations in the southern Sierra Nevada and the Mojave and Sonoran deserts. The leaves are generally but not always covered in fine hairs, giving them a velvety feel.

ID Tips: deciduous; leaves compound, leaflets 5-7, velvety on young twigs; seed pods (samaras) appearing in spring

California Big Tree Registry

The California Big Tree Registry recognizes, documents, and curates state and national champion tree records in California. The program receives big tree nominations for native and cultivated species in various environments—from national forests to historic urban areas—from people interested in trees. Recognizing these remarkable individuals brings pride to the communities surrounding the champion trees and helps residents of California further value and appreciate the environmental impact and societal benefits of trees.

Measuring a Big Tree

Trees are measured using a points formula to determine whether a tree is a champion: trunk circumference in inches + height in feet + ¼ of the average crown spread in feet = total point score. The individual within each species with the greatest point score is the champion.

Visit the California Big Tree Registry and learn about over 200 champions!

Champion *Populus fremontii*

Circumference (INCHES)

Height (FEET)

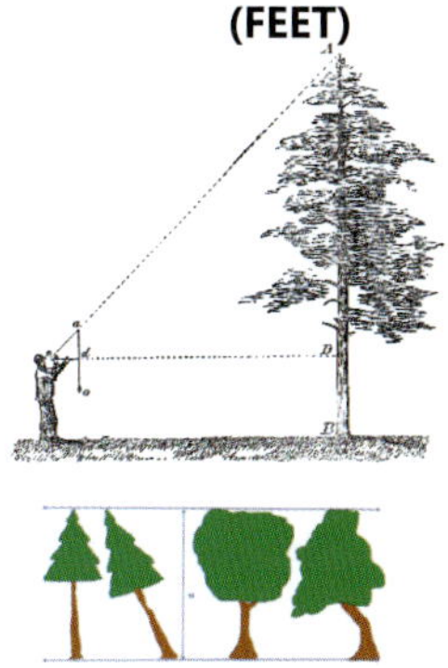

Crown Spread (FEET)

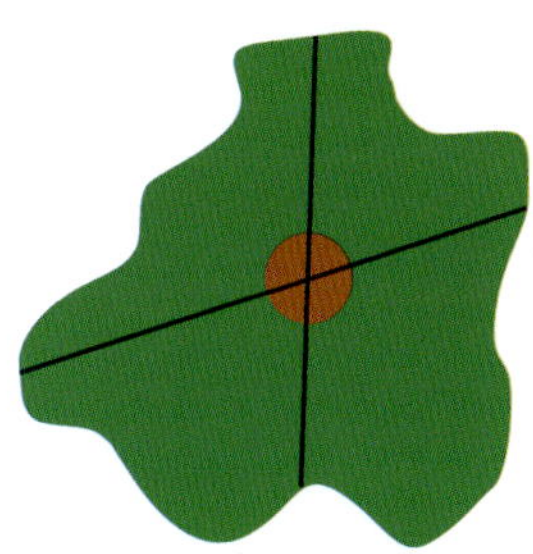

Average distance from one side of the canopy to the other

______ Inches + ______ Feet + ¼(______) Feet = Total Points

Sycamore Family

The Platanaceae is a small family with 12 species in the single genus, *Platanus*. These trees are recognized for their towering height, broad canopies, and distinctive bark that peels in patches to reveal smooth, mottled surfaces. In California, the California sycamore (*Platanus racemosa*) is a prominent native species, thriving in riparian habitats.

Platanus racemosa in the Tehachapi Mountains, Kern County

CALIFORNIA SYCAMORE

Platanus racemosa

PLATANACEAE

Height: to 110′
DBH: to 36″
Leaves: to 10″
Fruit: 0.5″-1.5″

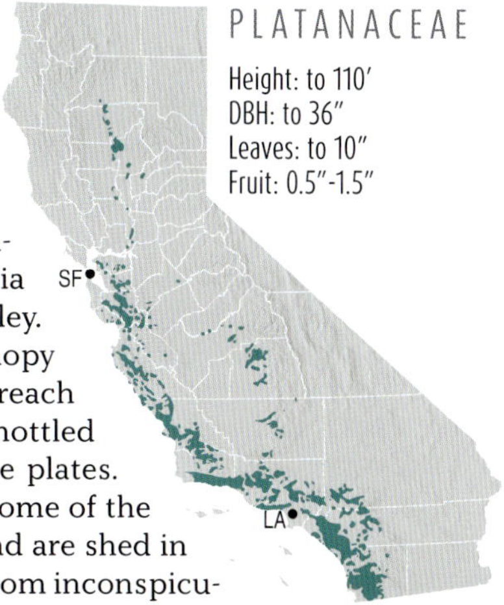

California sycamore ranges from northwestern Baja California to the northern Sacramento Valley. It grows in the canyons and waterways of western California and along the rivers of the Central Valley. These trees, with their spreading canopy of gnarled and crooked branches, can reach over 100 feet in height. The white bark, mottled with patches of gray, peels away in large plates. The velvety, palmately lobed leaves are some of the largest made by any tree in California and are shed in winter. Spherical fruit clusters develop from inconspicuous, maroon, early-spring flowers.

ID Tips: bark mottled, peeling in patches with smooth, pale inner bark; leaves palmate, 3-5-lobed; fruit clusters spherical, pendent

Rose Family

The Rosaceae is one of the largest and most diverse plant families in the world. It includes over 2,500 species spanning trees, shrubs, and herbaceous plants. Globally, it is renowned for its economic importance, with cultivated members such as apples (*Malus*) and cherries (*Prunus*). In North America, the Rosaceae includes native species like serviceberries (*Amelanchier* spp.), wild roses (*Rosa* spp.), and chokecherries (*Prunus virginiana).* In California, the Rosaceae is represented by a diverse array of native plants, including shrubs like California wild rose (*Rosa californica*) and toyon (*Heteromeles arbutifolia*). These species are integral to California's chaparral, oak woodland, and riparian ecosystems, offering nectar, berries, and shelter for pollinators, birds, and other animals. There is only one native tree.

Island ironwood is a rare and fascinating tree endemic to the Channel Islands. Fossil evidence reveals that the genus *Lyonothamnus* has ancient origins with relatives dating back to the mid-Tertiary period, over 23 million years ago, when the climate of western North America was much wetter and supported expansive forests. Over millions of years, as the region became more arid, most members of this genus went extinct, leaving *Lyonothamnus floribundus* on California's Channel Islands as a relic of a bygone era—it is, in essence, a living fossil!

A stand of *Lyonothamnus floribundus* on Santa Rosa Island, Santa Barbara County

ISLAND IRONWOOD

Lyonothamnus floribundus

ROSACEAE

Height: to 50′
DBH: to 15″
Leaves: 4″-6″
Flowers: 0.3″-0.5″

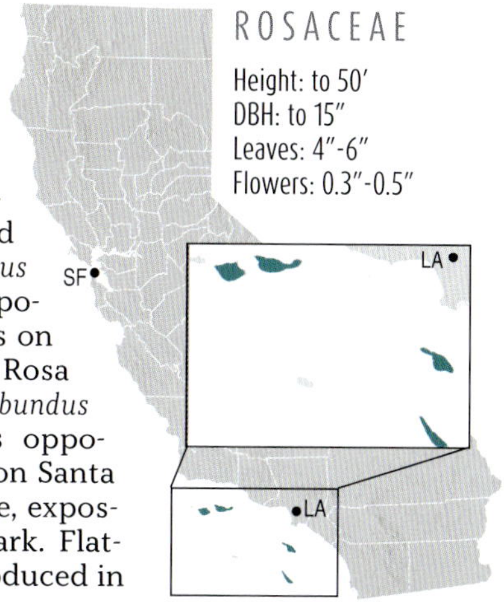

Island ironwood, though cultivated widely, occurs natively only on the California Channel Islands. The widespread horticultural subspecies *aspleniifolius* (Santa Cruz Island ironwood), has opposite, fern-like, divided leaves and occurs on San Clemente, Santa Cruz, and Santa Rosa islands. In contrast, subspecies *floribundus* (Santa Catalina Island ironwood) has opposite, undivided leaves and occurs only on Santa Catalina Island. The bark peels with age, exposing smooth, cinnamon-colored new bark. Flat-topped clusters of white flowers are produced in spring and linger on the tree for months.

floribundus

aspleniifolius

ID Tips: bark peeling in strips, reddish-brown; leaves opposite, simple or compound; clusters of small white flowers in late spring

Willow Family

The Salicaceae includes around 55 genera and over 1,200 species worldwide, with trees and shrubs distributed in temperate and tropical regions. In temperate regions its most notable members are poplars, aspens, cottonwoods (Populus spp.) and willows (Salix spp.). Globally, its most notable members are poplars, aspens, and cottonwoods (in the genus *Populus*), and willows (in the genus *Salix*).

The 40 or so species of aspen, cottonwood, and poplar are all in the genus *Populus*. Poplars are not in California where, generally, aspens are high-elevation trees and cottonwoods occur along waterways. The four species native to California are deciduous, fast-growing, and generally found along rivers, streams, or wherever the soil remains wet. Separate male and female individuals produce long, drooping clusters of inconspicuous, greenish-yellow, wind-pollinated flowers. Female trees form capsules that crack open in summer, dispersing seeds that ride in the wind on cottony fluff.

Of the 400 willow species worldwide, approximately 50 are native to California. Most of these are shrubs, such as California's most widely distributed willow, the arroyo willow (*S. lasiolepis*). However, three species in California mature into proper trees. Willows are fast-growing, deciduous plants that often dominate waterways and other wet habitats. Their flowers appear in elongated clusters on separate male and female individuals in winter, usually before the new leaves emerge. Fuzzy masses of hairy seeds disperse great distances on the wind and in flowing water.

Populus tremuloides in the fall are spectacular, Warner Mountains, Modoc County

NARROWLEAF COTTONWOOD

Populus angustifolia

SALICACEAE

Height: to 45′
DBH: to 12″
Leaves: 1″-4″
Fruit: 0.25″-0.4″

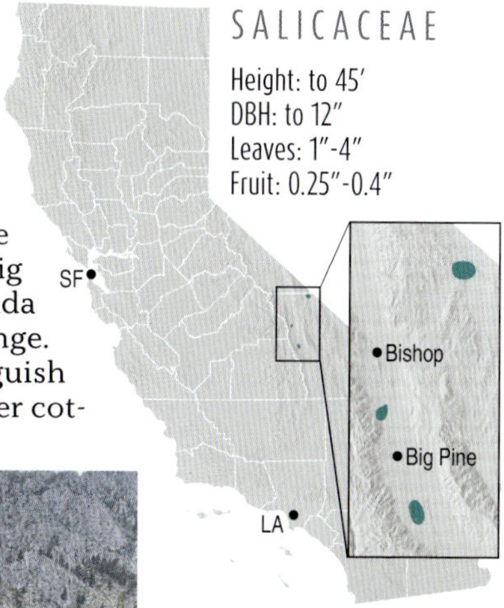

Narrow-leaved cottonwood is the most common cottonwood in the Rocky Mountains, yet is rare in California. The few populations in rocky ravines near Big Pine on the east side of the Sierra Nevada represent the western edge of its range. Narrow, spearhead-shaped leaves distinguish narrow-leaved cottonwood from the other cottonwoods in California.

ID Tips: trunk slender; bark smooth, yellowish-green when young; leaves narrow, lance-shaped, edges finely serrated

FREMONT COTTONWOOD

Populus fremontii

SALICACEAE

Height: to 110′
DBH: to 50″
Leaves: 1.2″-3″
Fruit: 0.2″-0.6″

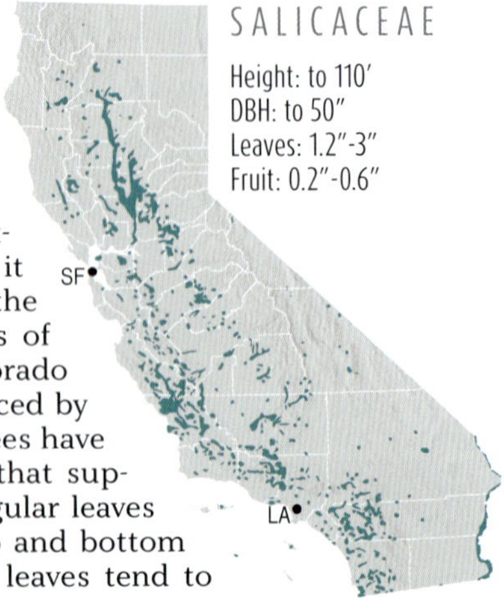

Fremont cottonwood grows along waterways throughout the southwestern U.S., from Texas through northern Mexico to coastal California. In California, it has a spotty distribution throughout the interior, especially along riparian areas of the Central Valley and along the Colorado River. In the coastal fog belt, it is replaced by black cottonwood (*P. trichocarpa*). Old trees have stout trunks covered in fissured bark that support a wide-spreading canopy of triangular leaves with coarsely scalloped edges. The top and bottom leaf surfaces are very similar, and the leaves tend to hang vertically.

ID Tips: crown broad, open; bark smooth when young, becoming furrowed; leaves heart-shaped or triangular, edges coarsely toothed

FLOWERING TREES

BLACK COTTONWOOD

Populus trichocarpa

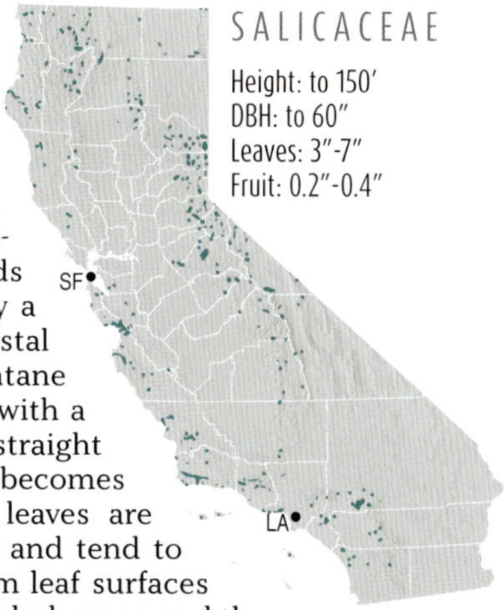

SALICACEAE

Height: to 150′
DBH: to 60″
Leaves: 3″-7″
Fruit: 0.2″-0.4″

Black cottonwood covers large areas of western North America from northern Baja California to Alaska, growing in riparian and moist bottomland woodlands and forests. In California it is primarily a tree of watercourses, particularly in coastal areas and into mid-elevations of montane areas. It can exceed 100 feet in height, with a bright green crown supported by a straight trunk covered in whitish bark that becomes fissured with age. Black cottonwood leaves are broadly triangular and finely scalloped and tend to spread horizontally. The top and bottom leaf surfaces are strongly contrasting, with the upper dark green and the lower dull and much paler.

ID Tips: bark furrows dark, rough; leaves lance-shaped, edges finely toothed, triangular; undersides dull and much paler

FLOWERING TREES

QUAKING ASPEN

Populus tremuloides

SALICACEAE

Height: to 60′
DBH: to 20″
Leaves: 1.25″-2.5″

Quaking aspen is North America's most widely dispersed tree species. It grows from the northeastern U.S. across Canada to Alaska, and southward through the Rockies into Mexico. In California, where it occurs primarily on the margins of Sierra Nevada and Cascade meadows. Across its range it sprouts from roots, forming large clonal colonies. Quaking aspen is a fast-growing, slender tree with smooth, whitish-green bark and leaves that shimmer and quake in the slightest breeze. The golden hues of its October foliage make this California's best species for fall color displays. Because of fire-suppression, some populations of quaking aspen have been replaced by coniferous forests.

ID Tips: Bark white or light gray, marked by black scars where lower branches are naturally self-pruned, smooth; leaves round-triangular

FLOWERING TREES

BLACK WILLOW

Salix gooddingii

SALICACEAE

Height: to 90′
DBH: to 15″
Leaves: to 5″
Female Catkins: to 3″

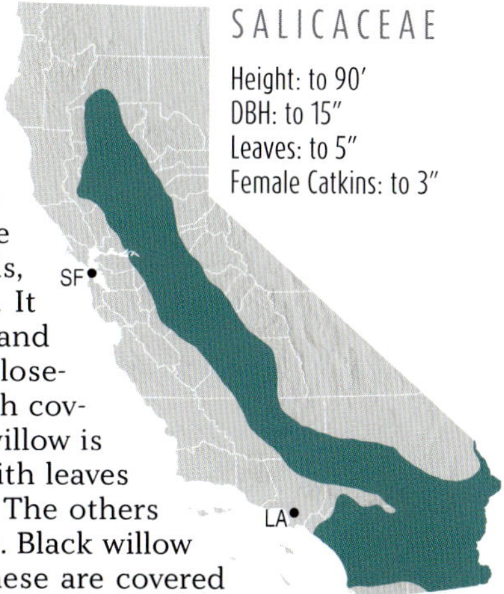

The black willow, the largest willow in the state, grows along riverbanks and in marshes and other wet areas in the Central Valley, Sierra Nevada foothills, and southeastern deserts of California. It extends eastward into Nevada, Arizona, and Texas, and southward into Mexico. It is closely related to black willow (*S. nigra*), which covers most of the eastern U.S. The black willow is the only tree-like willow in California with leaves that are similar in color on both sides. The others are lighter in color with dull undersides. Black willow often has multiple, crowded trunks. These are covered in rough, dark bark and support a rounded crown of linear leaves.

ID Tips: deciduous; leaves long, narrow, lance-shaped, both sides bright green, edges often finely serrated; wet areas

FLOWERING TREES

RED WILLOW

Salix laevigata

SALICACEAE

Height: to 45′
DBH: to 10″
Leaves: 4″-5″
Female Catkins: 1″-3″

Red willow occurs in gravelly and sandy areas along streambanks throughout western California and the Sierra Nevada foothills. Outside the state, it occasionally grows in small populations in Nevada, Oregon, Arizona, New Mexico, and Baja California. It has dark, grayish-brown bark on slender, twisted trunks supporting an irregular crown. Flexible, reddish-maroon twigs bear lance-shaped to oval leaves that are shiny and darker green on the upper surface and dull, bluish-green below.

ID Tips: young branches reddish; leaves narrow, lance-shaped, tops shiny, undersides duller; near water sources

PACIFIC WILLOW

Salix lasiandra

SALICACEAE

Height: to 50′
DBH: to 14″
Leaves: 1.5″-6.5″
Female Catkins: 0.5″-3.5″

Pacific willow grows throughout western California and the Sierra Nevada foothills and north to Alaska. It is distinguishable from other willows by the warty glands present where the leafstalk meets the blade. It has an upright, asymmetric crown with a short, dark trunk and long branches bearing yellowish-brown twigs and yellowish-green leaves that are shiny on the upper surface and whitish below.

ID Tips: shrub, small tree; twigs bright yellowish-brown, warty glands at base of each leaf stalk; leaves long, narrow, shiny above, finely toothed

FLOWERING TREES

Soapberry Family

The Sapindaceae has over 130 genera and over 1,800 species distributed worldwide, particularly in tropical and subtropical regions. Globally, the family includes economically and ecologically important species such as the soapberry (*Sapindus* spp.), lychee (*Litchi chinensis*), longan (*Dimocarpus longan*), horse-chestnut (*Aesculus hippocastanum*), and many maples (*Acer* spp.). In California, native members of the Sapindaceae include the maples (*Acer* spp.) with four species, two of which are trees, and California buckeye (*Aesculus californica*).

One of the world's largest *Acer macrophyllum*, Humboldt County

BIGLEAF MAPLE

Acer macrophyllum

SAPINDACEAE

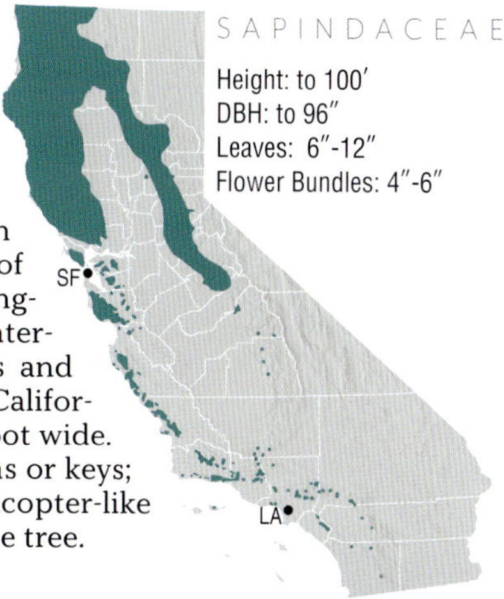

Height: to 100′
DBH: to 96″
Leaves: 6″-12″
Flower Bundles: 4″-6″

The 130 or so of the world's maple species grow almost entirely in temperate areas of the Northern Hemisphere. In California, we have four species, two of which are trees. The largest and most long-lived of these is bigleaf maple. This winter-deciduous tree grows in riparian areas and mixed-evergreen forests from southern California to Alaska. Its lobed leaves can be a foot wide. Maples have winged fruits called samaras or keys; these are formed in pairs that spin helicopter-like through the air when dispersing from the tree.

ID Tips: leaves hand-like, largest of any maple; major component of mid-elevation mixed-evergreen forests

FLOWERING TREES

BOX ELDER

Acer negundo

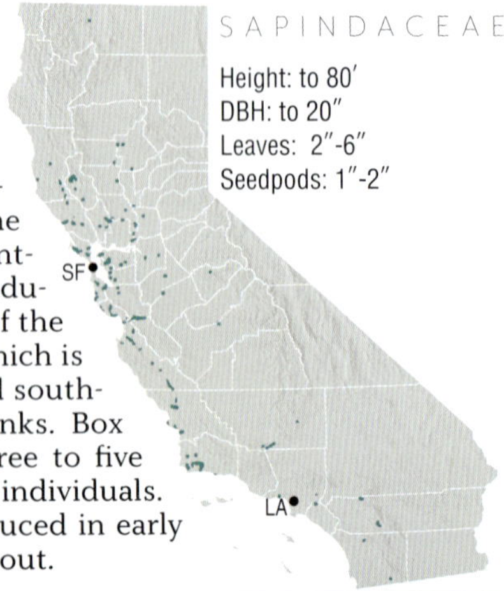

SAPINDACEAE

Height: to 80′
DBH: to 20″
Leaves: 2″-6″
Seedpods: 1″-2″

Over the past five to ten million years, during the formation of western California's Mediterranean-type climate, the state gradually became drier. Consequently, California shares very few of the deciduous hardwood trees typical of the rest of the continent. An exception is box elder, which is found throughout much of the U.S. and southern Canada, typically along streambanks. Box elders have compound leaves with three to five leaflets and separate male and female individuals. Their wind-pollinated flowers are produced in early spring before or as the trees are leafing out.

ID Tips: twigs smooth, green (purplish); leaves compound with three leaflets; seeds winged (samaras), hanging in clusters

CALIFORNIA BUCKEYE

Aesculus californica

SAPINDACEAE

Height: to 40′
DBH: to 20″
Leaves: 2.5″-6.5″
Flower Bundles: 6″-8″

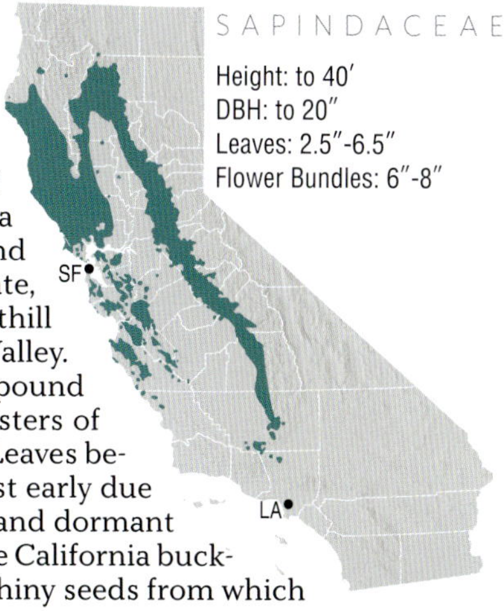

Aesculus is a genus of 15 species of trees and shrubs from temperate regions of North America and Eurasia. California buckeyes are found only in California and are the sole native buckeyes in the state, growing in the Coast Ranges and foothill woodlands surrounding the Central Valley. They have opposite, palmately compound leaves. In spring, they produce long clusters of sweetly scented, pinkish-white flowers. Leaves begin to grow earlier than most but are lost early due to summer drought. Trees stay leafless and dormant until the following spring. All parts of the California buckeye are poisonous, including the large, shiny seeds from which the plant gets its common name.

ID Tips: deciduous, small-statured; leaves palmately compound with 5-7 leaflets; flowers large, fragrant; fruit capsules pear-shaped

Large and Iconic Shrubs (almost trees)

Many of California's most iconic plants are shrubs. Some of these are listed below. See, What is a tree on page 14.

Scientific Name	Family	Common Name
Acer circinatum and *glabrum*	Sapindaceae	VINE and ROCKY MOUNTAIN MAPLE
Adenostoma sparsifolium	Rosaceae	RED SHANKS
Alnus viridis	Betulaceae	GREEN ALDER
Amelanchier utahensis	Rosaceae	UTAH SERVICE-BERRY
Arctostaphylos glauca, manzanita, and other species	Ericaceae	BIG BERRY MANZANITA, COMMON MANZANITA, and other species
Artemisia tridentata	Asteraceae	GREAT BASIN SAGEBRUSH
Betula occidentalis	Betulaceae	WATER BIRCH
Bursera microphylla	Burseraceae	ELEPHANT TREE
Calycanthus occidentalis	Calycanthaceae	SPICEBUSH
Ceanothus cuneatus, thyrsiflorus, and other species	Rhamnaceae	BUCKBRUSH, BLUE BLOSSOM, and other species
Cephalanthus occidentalis	Rosaceae	CALIFORNIA BUTTON WILLOW
Cercis occidentalis	Fabaceae	WESTERN REDBUD
Cercocarpus spp.	Rosaceae	MOUNTAIN-MAHOGANY
Chrysolepis sempervirens	Fagaceae	BUSH CHINQUAPIN
Comarostaphylis diversifolia	Ericaceae	SUMMER HOLLY
Corylus cornuta subsp. *californica*	Betulaceae	CALIFORNIA HAZEL
Dendromecon rigida	Papaveraceae	BUSH POPPY
Ericameria arborescens	Asteraceae	GOLDEN-FLEECE
Frangula californica and *purshiana*	Rhamnaceae	COFFEEBERRY, CASCARA
Fraxinus anomala, dipetala, and *parryi*	Oleaceae	SINGLE-LEAF ASH, CALIFORNIA ASH, CHAPARRAL ASH

ROCKY MTN. MAPLE - *Acer glabrum*

RED SHANKS - *Adenostoma sparsifolium*

G.B. SAGEBRUSH - *Artemisia tridentata*

SPICEBUSH - *Calycanthus occidentalis*

CALIFORNIA LILAC - *Ceanothus* sp.

WESTERN REDBUD - *Cercis occidentalis*

Large and Iconic Shrubs (almost trees)

Fremontodendron californicum and *mexicanum*	Malvaceae	FLANNELBUSH, MEXICAN FLANNELBUSH
Garrya elliptica and *veatchii*	Garryaceae	SILK TASSEL BUSHES
Heteromeles arbutifolia	Rosaceae	TOYON
Holodiscus discolor	Rosaceae	OCEANSPRAY
Juniperus californica and *occidentalis*	Cupressaceae	CALIFORNIA, WESTERN JUNIPERS
Malosma laurina	Anacardiaceae	LAUREL SUMAC
Morella californica	Myricaceae	WAX MYRTLE
Oemleria cerasiformis	Rosaceae	OSOBERRY
Parkinsonia microphylla	Fabaceae	LITTLE-LEAF PALO VERDE
Prunus emarginata, ilicifolia, and *lyonii*	Rosaceae	BITTER, HOLLY-LEAFED, CATALINA CHERRIES
Psorothamnus spinosus	Fabaceae	SMOKE TREE
Ptelea crenulata	Rutaceae	HOP TREE
Quercus spp.	Fagaceae	11 species of Ca. scrub oaks
Rhamnus ilicifolia and *crocea*	Rhamnaceae	HOLLYLEAF REDBERRY, REDBERRY
Rhododendron macrophyllum	Ericaceae	CALIFORNIA RHODODENDRON
Rhus integrifolia and *ovata*	Anacardiaceae	LEMONADE BERRY, SUGAR BUSH
Salix lasiolepis and others	Salicaceae	ARROYO WILLOW, other willow species
Sambucus mexicana and *racemosa*	Caprifoliaceae	BLACK, RED ELDERBERRY
Senegalia greggii	Fabaceae	CATCLAW ACACIA
Staphylea bolanderi	Staphyleaceae	SIERRA BLADDERNUT
Styrax redivivus	Styracaceae	SNOWDROP BUSH
Vachellia farnesiana	Fabaceae	SWEET ACACIA

MTN. MAHOGANY - *Cercocarpus* spp.

BUSH POPPY - *Dendromecon rigida*

CASCARA - *Frangula purshiana*

FLANNELBUSH - *F. californicum*

OCEANSPRAY - *Holodiscus discolor*

CA. JUNIPER - *Juniperus californica*

Large and Iconic Shrubs (almost trees)

WAX MYRTLE - *Morella californica*

OSOBERRY - *Oemleria cerasiformis*

BITTER CHERRY - *Prunus emarginata*

MULLER'S OAK - *Q. cornelius-mulleri*

LEATHER OAK - *Quercus durata*

SADLER OAK - *Quercus sadleriana*

LEMONADE BERRY - *Rhus integrifolia*

ARROYO WILLOW - *Salix lasiolepis*

RED ELDERBERRY - *Sambucus racemosa*

CATCLAW ACACIA - *Senegalia greggii*

SIERRA BLADDERNUT - *Staphylea bolanderi*

SNOWDROP BUSH - *Styrax redivivus*

Non-native Trees

California is home to around 6,000 native species of plants. Since the first European incursions into the state, countless others have been brought to California, purposefully or accidentally. Now, over 1,000 of these non-native species grow and reproduce outside of human cultivation. Some are invasive and can spread into wild habitats, displace native species, and alter entire ecosystems.

Most invasive plants in California are non-woody. Invasive tree species are rare in California, especially outside watercourses, as most of the state is too dry for non-native trees to become established.

The California Invasive Plant Council (Cal-IPC) maintains an inventory of invasive plants based on ecological impacts and a "Watch List" of species that may become invasive. As of 2025, the list contains 341 species, 34 of which are shrubs and 29 of which are trees.

Below is the list of the non-native trees that are invasive or weedy in some specific areas of California wildlands. None of these species are invasive everywhere they grow, and most only invade watercourses.

SCIENTIFIC NAME	COMMON NAME	FAMILY	Cal-IPC Rating
Acacia melanoxylon	BLACKWOOD ACACIA	FABACEAE	Limited
Acacia dealbata	SILVER WATTLE	FABACEAE	Moderate
Crataegus monogyna	COMMON HAWTHORN	ROSACEAE	Limited
Elaeagnus angustifolia	RUSSIAN OLIVE	ELAEAGNACEAE	Moderate
Eucalyptus camaldulensis	RED GUM	MYRTACEAE	Limited
Eucalyptus globulus	BLUE GUM	MYRTACEAE	Limited
Ficus carica	EDIBLE FIG	MORACEAE	Moderate
Grevillea robusta	SILK OAK TREE	PROTEACEAE	Limited
Ligustrum lucidum	GLOSSY PRIVET	OLEACEAE	Limited
Maytenus boaria	MAYTEN TREE	CELASTRACEAE	Limited
Myoporum laetum	MYOPORUM	MYOPORACEAE	Moderate
Olea europaea	OLIVE	OLEACEAE	Limited
Phoenix canariensis	CANARY ISLAND DATE PALM	ARECACEAE	Limited
Prunus cerasifera	PURPLE-LEAF PLUM	ROSACEAE	Limited
Robinia pseudoacacia	BLACK LOCUST	FABACEAE	Limited
Schinus molle	PEPPER TREE	ANACARDIACEAE	Limited

SCIENTIFIC NAME	COMMON NAME	FAMILY	Cal-IPC Rating
Schinus terebinthifolia	BRAZILIAN PEPPER	ANACARDIACEAE	Moderate
Tamarix aphylla	ATHEL TREE	TAMARICACEAE	Limited
Triadica sebifera	CHINESE TALLOW TREE	EUPHORBIACEAE	Moderate
Washingtonia robusta	MEXICAN FAN PALM	ARECACEAE	Moderate

BLACKWOOD ACACIA - *A. melanoxylon*

RED GUM - *Eucalyptus camaldulensis*

BLUE GUM - *Eucalyptus globulus*

SILK OAK TREE - *Grevillea robusta*

GLOSSY PRIVET - *Ligustrum lucidum*

OLIVE - *Olea europaea*

CANARY ISL. DATE PALM - *P. canariensis*

PURPLE-LEAF PLUM - *Prunus cerasifera*

BLACK LOCUST - *Robinia pseudoacacia*

ATHEL TREE - *Tamarix aphylla*

EDIBLE FIG - *Ficus carica*

COMMON HAWTHORN - *C. monogyna*

RUSSIAN OLIVE - *Elaeagnus angustifolia*

BRAZILIAN PEPPER - *S. terebinthifolia*

MEXICAN FAN PALM - *W. robusta*

CHINESE TALLOW TREE - *Triadica sebifera*

Southern California Tree Destinations

Los Padres National Forest (1)
Exceptional tree diversity along lowland rivers like the Sisquoc and Sespe to higher-elevation conifer forests on Pine Mountain.

Santa Monica Mountains (2)
Black walnuts, sycamores, alders, and more decorate the stunning canyons in this urban oasis. Watch for poison oak, too!

San Gabriel Mountains (3)
Drive Highway 2 from La Cañada to Wrightwood to enjoy dramatic views and the transition into the highly diverse conifer forests.

Lake Isabella (4)
Enjoy the Kern River country with lowland trees, and explore the mid-elevations for the rare Piute cypress.

Torrey Pines State N. R. (5)
Just north of San Diego, this quaint park protects one of the world's rarest pines with stunning views across the Pacific Ocean.

San Bernardino Mountains
From Big Bear Lake to the Pacific Crest Trail, there are endless adventures in the high country of this part of Transverse Ranges. (6)

San Jacinto Mountains
Take the tram 6,000 feet up from the Sonoran Desert floor to the high peaks of the mountains covered in conifer forests with views for days. (7)

Anza-Borrego Desert S. P.
This low-elevation park is the perfect winter and early spring destination, with Ca. fan palms and blue palo verde, among other trees. (8)

Joshua Tree National Park
The park's tree namesake is found in the western part while the Sonoran Desert defines the eastern half, with ironwood and mesquite. (9)

Mojave National Preserve
Visit the Kelso Dunes to see the heart of the park and dunes with desert-willow. There are also campgrounds and caves. (10)

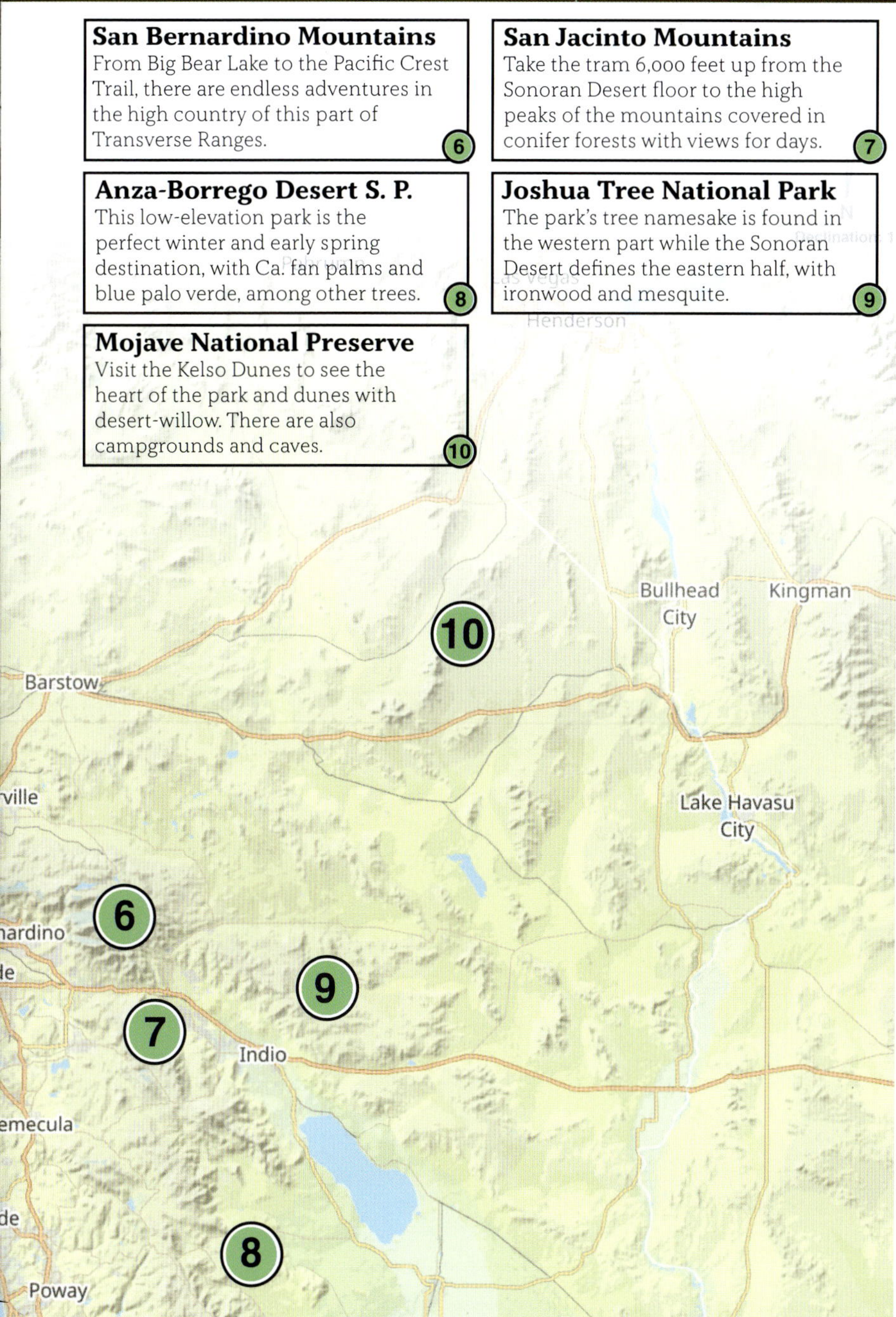

MORE TO EXPLORE

South-Central California Tree Destinations

Big Sur
See the rare Santa Lucia fir on Cone Peak, drive 101 to the southern range extension of redwood, and enjoy some of California's best scenery. 1

Point Lobos S. N. R.
Classic California coastline and some of the state's most iconic scenery with the rare Monterey cypress, Gowen cypress, and more. 2

Big Basin Redwoods S. P.
A variety of hikes through redwoods and more provide the opportunity to observe the post-fire response in coastal forests. 3

Pinnacles National Park
Beautiful park with California condors soaring and perching in ghost pines with unique rock formations all around. 4

San Benito Mountains
Unique Coast Range geology with higher elevation serpentine soils driving tree diversity including Jeffrey pine and ghost pine. 5

Giant Sequoia N. M.
Protecting over 500 square miles and half of the giant sequoia groves in existence, this monument has endless hiking and driving opportunities. 6

Sequoia-Kings Canyon N. P.
Roads will bring you to the mid-elevations but only trails can take you into the high country with foxtail pines and other unique subalpine species. 7

White Mountains
Drive to one of the most unique forests on Earth, then hike through the groves to appreciate the survival and lives of the world's oldest trees. 8

Death Valley National Park
From the lowest elevation in North America, across desert washes and canyons, and to the summit of Telescope Peak, adventure awaits. 9

Ash Meadows N. W. R.
Just across the California border in Nevada, this refuge has crystal pools containing rare desert pupfish and lined with velvet ash. 10

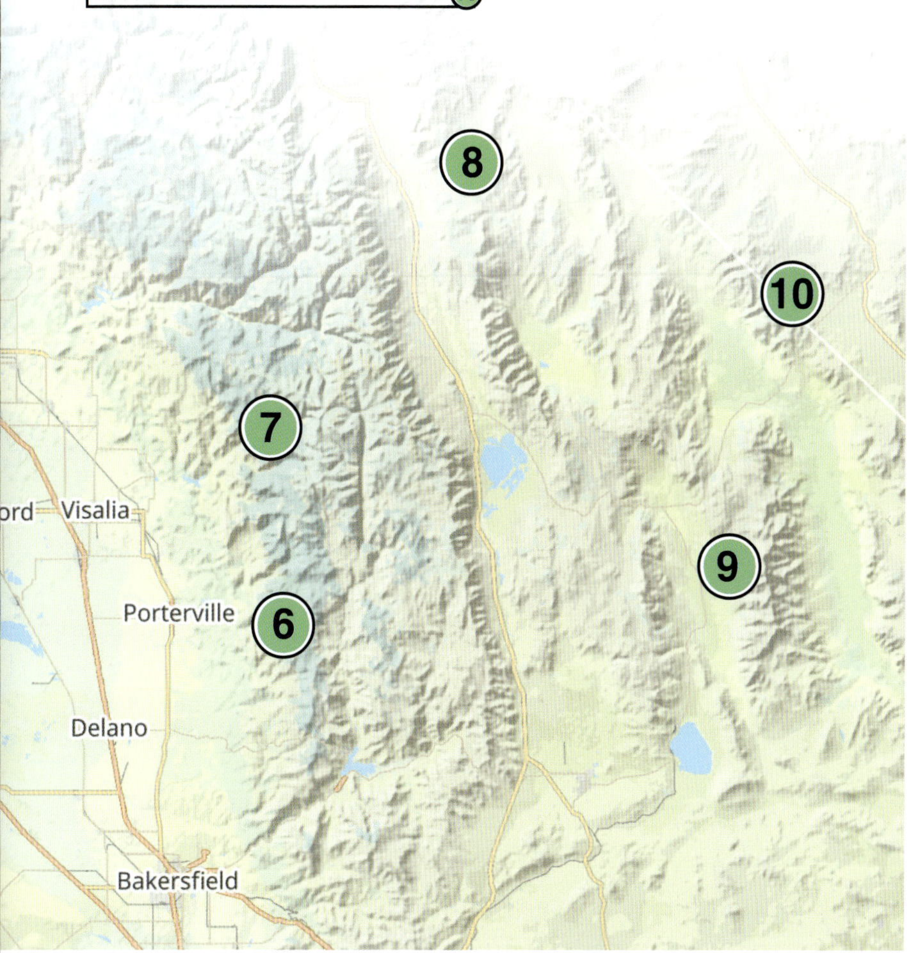

North Central California Tree Destinations

Muir Woods N. M.
Popular destination for Bay Area travelers to enjoy redwoods in the lower canyons but don't miss the Sargent cypresses on Mount Tam! 1

Point Reyes Nat. Seashore
In addition to dramatic spring wildflowers, watch for Douglas-fir, bishop pine, coast live oak, and bay laurel. 2

Robert Louis Stevenson S.P.
Hike Mount St. Helena for sugar pines, stay in the canyons for California nutmegs, all with some of the highest tree diversity in the state. 3

Jug Handle State N. R.
Explore coastal forests including the unusual stands of the North Coast endemic pygmy cypress mixed with madrone and bishop pine. 4

Berryessa Snow Mtn. N. M.
Explore the transition from chaparral to oak woodland to red fir forest on Snow Mountain in this stunning 330,780-acre monument. 5

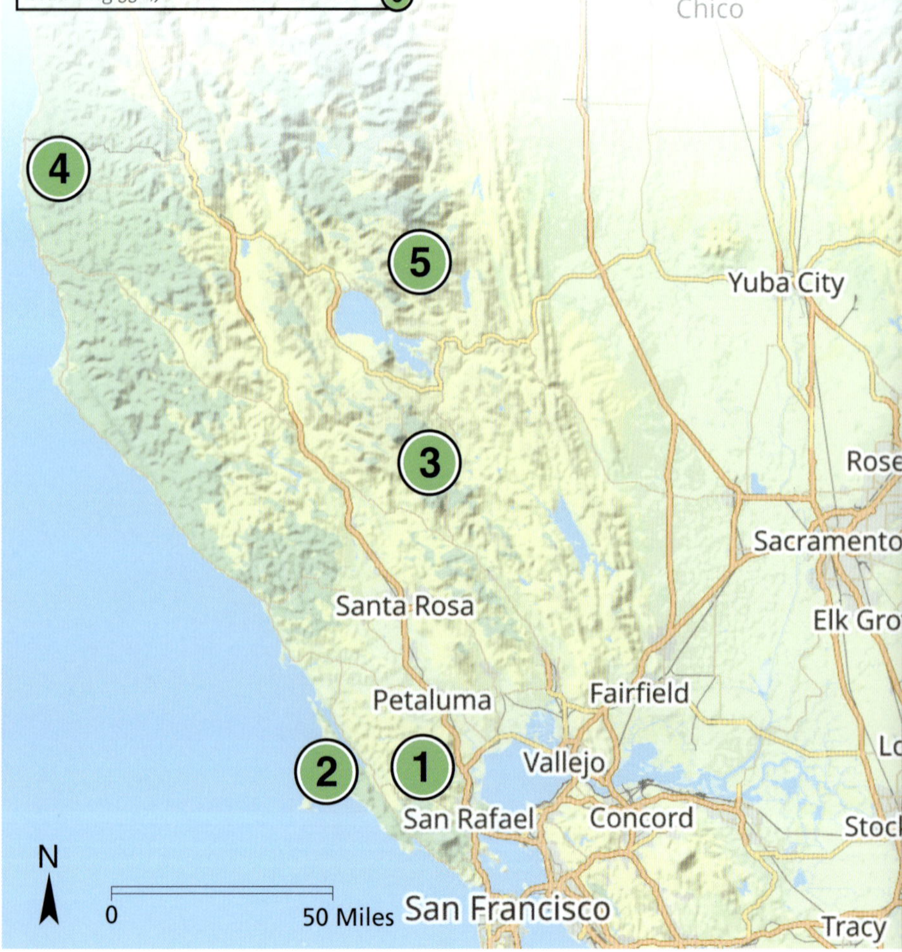

Yosemite National Park
From the Valley floor to the high country, the rocks will inspire but the plant life will keep you intrigued. Be sure to get on a trail somewhere. 6

Mono Lake - Eastern Sierra
As water drains from the high Sierra to forested canyons of cottonwoods and aspen below the conifer-covered peaks above. 7

Calaveras Big Trees S. P.
A prime example of a mixed-conifer forest, with giant sequoias among ponderosa pines, sugar pines, incense-cedars, Pacific yews, and white fir. 8

Desolation Wilderness
Amazing access to high Sierra landscape of dramatic lake basins and peaks covered with western white and whitebark pines. 9

Lake Tahoe
This high-elevation environment provides ideal conditions for a variety of coniferous trees like sugar, Jeffrey, and ponderosa pines. 10

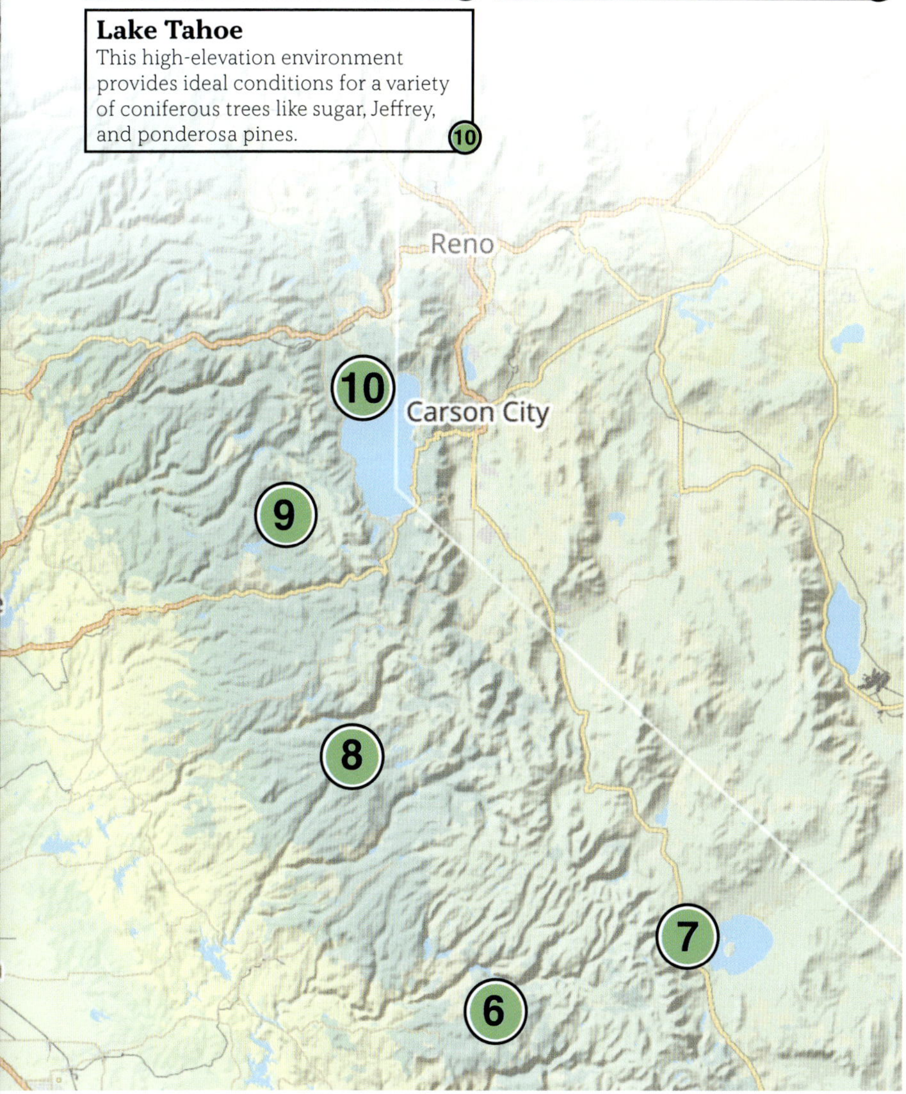

MORE TO EXPLORE

Northern California Tree Destinations

Humboldt Redwoods S. P.
One of California's largest state parks offers miles of trails through redwoods and upland hardwoods with tanoak and madrone. 1

Sue-meg State Park
Oceanside vistas and lonesome beaches compliment unusual coastal forests with the northern extension of bishop pine. 2

Prairie Cr. Redwoods NP/SP
Stunning forests that mix redwoods with species common farther north like western hemlock, Sitka spruce, and western redcedar. 3

Smith River N. R. A.
High rainfall sculpts the emerald Smith River and supports amazing tree diversity including Port Orford-cedar and giant chinquapin. 4

Klamath Mountains
One of the most biodiverse temperate mountain ranges holds endemic trees like Brewer spruce and other regional rarities like yellow-cedar. 5

Mount Shasta
One of California's highest mountains holds glaciers and deep winter snow. Drive to Bunny Flat to enjoy subalpine forests with hemlocks and firs. (6)

Lassen Volcanic N. P.
Drive the park and explore a variety of natural wonders like whitebark pine near Helen Lake and bubbling mud pots with red firs and mtn. hemlocks. (7)

Ahjumawi Lava Springs S. P.
Plants include western juniper, Oregon white oak, ponderosa pine, and shrubs like bitterbrush, mahogany, and ceanothus. (8)

Lava Beds National Mon.
Explore the edge of the Great Basin with underground lava tubes topped with a wide range of plants includng ferns and western junipers. (9)

Modoc County
Stunning vistas await in Great Basin country. Don't miss the Warner Mountains with tree species like quaking aspen and incense-cedar. (10)

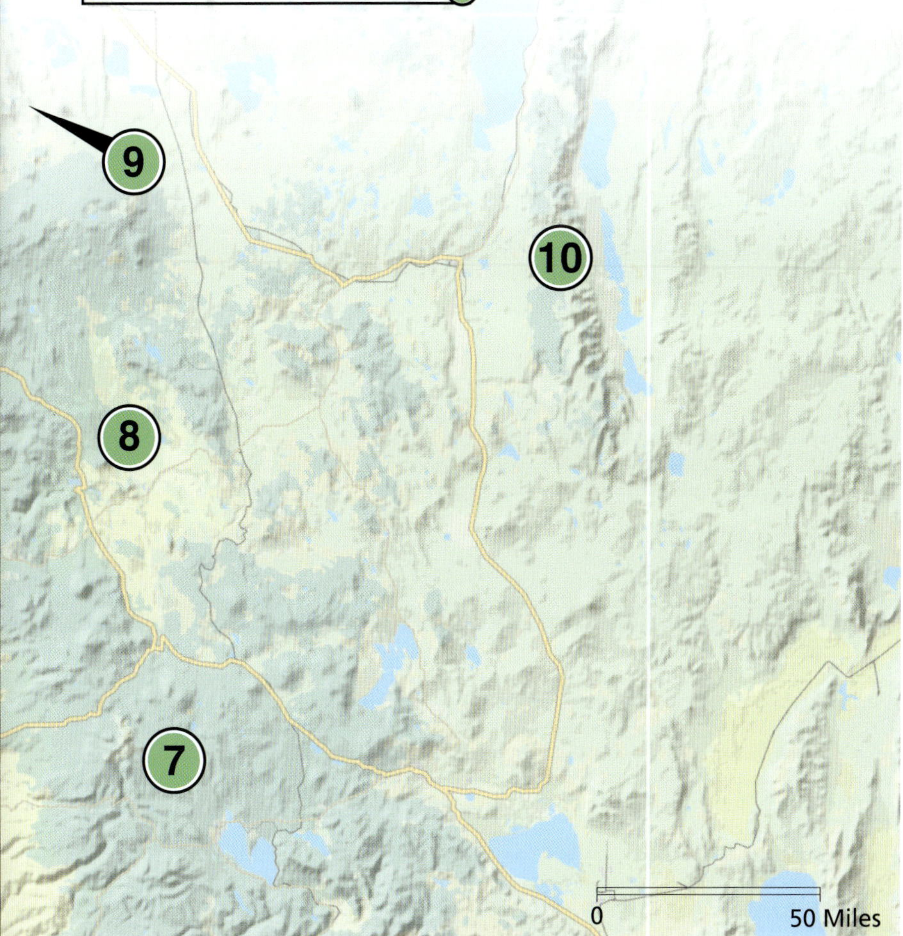

Resources

Bakker, E. S. (1985). *An Island Called California: An ecological introduction to its natural communities* (2nd ed.). University of California Press.

Griffin, J. R., & Critchfield, W. B. (1972). *The Distribution of Forest Trees in California*. Res. Paper PSW-RP-82. Pacific Southwest Forest and Range Experiment Station, U.S. Department of Agriculture, Forest Service.

Jepson, W. L. (1923). *The Trees of California* (2nd ed.). Associated Students Store, University of California, Berkeley.

Baldwin, B. G., Goldman, D. H., Keil, D. J., Patterson, R., & Rosatti, T. J. (eds.). (2012). *The Jepson manual: Vascular plants of California* (2nd ed.). University of California Press.

Kauffmann, M. E. (2023). *Conifer Country: A Natural History and Hiking Guide to the Conifers of the Klamath Mountain Region* (2nd ed.). Backcountry Press.

Kauffmann, M. E. (2024). *Conifers of the Pacific Slope: A field guide to the conifers of California, Oregon, and Washington* (2nd ed.). Backcountry Press.

Lanner, R. M. (1999). *Conifers of California*. Cachuma Press.

Little, E. L. (1971). *Atlas of United States trees (Vol. 1): Conifers and important hardwoods*. U.S. Dept. of Agriculture, Forest Service. https://doi.org/10.5962/bhl.title.130546

McMinn, H., & Maino, E. (1935). *An Illustrated Manual of Pacific Coast Trees* (1st ed.). University of California Press.

Pavlik, B. M., Muick, P. C., Johnson, S. G., & Popper, M. (1991). *Oaks of California*. Cachuma Press and the California Oak Foundation.

Ritter, Matt. (2018). *California Plants: A Guide to Our Iconic Flora*. Pacific Street Publishing.

Ritter, M. (2022). *A Californian's Guide to the Trees Among Us* (Expanded and updated). Heyday.

Rundel, Phil, R. Gustafson, M. Kauffmann (2022). *California Desert Plants: Ecology and Diversity*. Backcountry Press.

Sawyer, J. O., Keeler-Wolf, T., & Evens, J. (2008). *A Manual of California Vegetation* (2nd ed.). California Native Plant Society Press.

Schoenherr, A. A. (2017). *A Natural History of California* (2nd ed.). University of California Press.

Sudworth, G. B. (1908). *Forest Trees of the Pacific Slope*. U.S. Dept. of Agriculture, Forest Service.

Van Pelt, R. (2001). *Forest Giants of the Pacific Coast*. Global Forest Society in association with University of Washington Press, Seattle.

Watson, F. D., & Eckenwalder, J. E. (1993). *Cupressaceae. In: Flora of North America Editorial Committee (eds.)*. 22+ vols. New York and Oxford. Vol. 2.

Acknowledgments

MATT RITTER: My deepest appreciation goes to my field trip companions, friends, and colleagues Jenn Yost, May Reid-Marr, Reed Kenny, Andrew Fricker, Cami Pawlak, Natalie Love, Kieran Althaus, Jared Farmer, Jason Johns, Taylor Crow, Gage Willey, and Annie Meeder, who've accompanied me on many tree adventures. Thank you to Randy Baldwin, Dylan Hannon, Don Hodel, Jason Dewees, Kathy Musial, Nicole Cavender, Ken Greby, Mike Simpson, Jon Rebman, Jim Henrich, Jim Mauseth, Francis Halle, Dave Muffly, Ian Kesterson, Walt Gault, Jim Clark, Nelda Matheny, and Dave Teuchler for the tree knowledge and inspiration. Thank you to Cami Pawlak for her cartographic skills. Really, this book couldn't have happened without her. Thank you to Jenn Yost, Dylan Neubauer, Dave Keil, and Sairus Patel for carefully reading early drafts of the manuscript. Thank you to Sarah Ritter as always for all her help and support.

Photographic Contributions

Zach Abbey* p. 180 (*Acer negundo,* both right photos)
Wendy Anderson* p. 131 (*Celtis reticulata*, bottom left)
Barbara Banfeild* p . 181 (*Aesculus californica*, bottom right)
Millie Basden* p. 180 (*Acer negundo,* bottom left)
Matt Berger* p. 139 (*S. pubescens,* top left and bottom photos), p. 152 (*Quercus lobata* bottom right), p. 167 (*Platanus racemosa*, top right), p. 190 (*Robinia pseudoacacia)*
Jeff Bisbee p. 69 (*A. amabilis,* seed cones), p. 70 (*A. bracteata,* seed cones)
Dave Bunnell* p. 59 (Bennett juniper)
Peter Chen* p. 190 (*Phoenix canariensis)*
Ryan Donnelly* p. 148 (*Quercus douglasii*, acorn)
Melanie Davis* p 128 (*Carnegiea gigantea*, flowers)
Chris Earle p. 25 (Laguna Mountains), p. 31 (*Sequoiadendron giganteum*), p. 46 (*Hesperocyparis abramsiana*, bottom left), p. 48 (*Hesperocyparis forbesii*, bottom two), p. 49 (*Hesperocyparis goveniana*, top left)
Kevin Floyd* p. 139 (*S. pubescens,* top right), p. 141 (*Parkinsonia florida*, top right)
Mary K. Hanson* p. 159 (*Juglans hindsii*, top right), p. 180 (*Acer negundo* top left), p. 191 (*Crataegus monogyna)*
Eric Alan Isaacson* p. 191 *(S. terebinthifolia)*
Gary Jahns* p. 152 (*Quercus lobata* bottom left)
David Keil p. 108 (*T. heterophylla* seed cones), p. 125 (*A. rhombifolia* bottom), p. 138 (*N. odorata* flower), p. 145 (*Q. agrifolia* top middle and bottom right), p. 146 (Q. chrysolepis top right), p. 164 (*F. latifolia* top right), p. 173 (*P. trichocarpa* bottom right), p. 175 (*S. gooddingii* top right), p. 177 (*S. lasiandra* bottom center)
Sam Kieschnick* p. 131 (*Celtis reticulata*, bottom right)
Matt Lavin* p. 158 (walnuts)
Steve Matson* p. 165 (*F. velutina,* top left and bottom right), p. 166 (*Platanus racemosa)*
Irina Mitine* p. 190 (*Olea europaea)*
Jim Morefield* P. 191 (*Elaeagnus angustifolia)*
Dylan Neubauer p. 23 (high elevation White Mountain forest)
Harrison Reiber* p. 160 (*Umbellularia californica*)
Abraham Sánchez Romero* p. 137 (*Olneya tesota*, three bottom photos)
Katja Schulz* p. 128 (*Carnegiea gigantea*, top left)
Adam Searcy* p. 177 (*Salix lasiandra,* top left)
Mike Splain pp. 10-11 (*Quercus lobata*)
Chloe and Trevor Van Loon* p. 187 (*Staphylea bolanderi)*
Bob Van Pelt p. 147 (Smither's Oak)
Lindsay Wood* p. 191 *(Triadica sebifera)*
Joergmlpts* p. 153 (*Quercus parvula* var. *shrevei*, two top left)

Acknowledgments

MICHAEL KAUFFMANN: Tree lovers are born from hands-on experiences. Mine started in high school when I was introduced to Virginia's trees by my teacher Charles Dubay. My parents help nurture that love as well. When I moved to California my boss Rick Mitchell and co-worker RT Hawke inspired me to learn more about California's trees. Once in Humboldt, John Sawyer challenged me solve Klamath conifer mysteries with other Cal Poly Humboldt mentors and colleagues like Steve Sillett, Marie Antoine, Lucy Kerhoulas, Erik Jules, and Rosemary Sherriff. Today, my family continues to inspire me with explorations and questions about trees—thanks Allison, Sylas, and Asa for tree-seeking adventures with me! I thank Matt Ritter for the vision for this project. Chris Earle offered solid draft reviews of this book. Thérèse Shere is an amazing indexer. Special thanks to Dylan Neubauer for her sharp edits. Photographic inspiration came from many.

Index

Note: Page numbers in **bold** refer to tree species profiles that include photos, range maps, and ID tips. Page numbers in *italics* refer to photos and captions.

MORE TO EXPLORE

About the Authors

MATT RITTER

Grew up in rural Mendocino County, California. After earning a bachelor's degree in microbiology from U.C. Santa Barbara, he attended U.C. San Diego for a Ph.D. in plant biology. He's a botany professor in the Biological Sciences Department at Cal Poly in San Luis Obispo, California, where he teaches plant biology, ecology, and field botany. He's the author of several books including *California Plants: A Guide to Our Iconic Flora*. He also wrote the best-selling guide to California's urban forest, *A Californian's Guide to the Trees Among Us* (Heyday, 2022). He recently won the National Outdoor Book Award for his children's book, *Something Wonderful*. He won the Cal Poly's Distinguished Teaching Award, the International Society of Arboriculture Award for Excellence in Education, and the Indie Book Award for his climate change novel *Rainwalkers*. He is the California Coordinator of the National Big Tree Registry and studies California's native plants and trees in the urban forest. He's an avid woodworker, mason, and gardener.

MICHAEL KAUFFMANN

Grew up in the eastern deciduous forests of Virginia and would stare out the window in the winter, hooting for barred owls and wonder...why are some of those trees still green? He nurtured his evergreen conifer fascinations in the Appalachian Mountains with a biology degree from Virginia Tech. Upon graduation, found a job teaching outdoor education in California. Sans family, alone in the West, he began to hang out with giant sequoias and foxtail pines on weekends. The temperate rainforest eventually called him north to Humboldt County where he earned a biology Master's degree from Cal Poly Humboldt for his first book *Conifer Country*. He now lives in Kneeland, California, with his wife Allison, and his sons, Sylas and Asa. He has also authored *Conifers of the Pacific Slope* and co-authored *The Klamath Mountains: A Natural History*, *Field Guide to Manzanitas*, and *California Desert Plants*. Besides writing, he works as a devoted educator, is the founder and president of the Bigfoot Trail Alliance, and is a research plant ecologist.